*Foundations of Rational Choice
under Risk*

Foundations of Rational Choice Under Risk

PAUL ANAND

CLARENDON PRESS • OXFORD

1993

Oxford University Press, Walton Street, Oxford OX2 6DP
Oxford   New York   Toronto
Delhi   Bombay   Calcutta   Madras   Karachi
Kuala Lumpur   Singapore   Hong Kong   Tokyo
Nairobi   Dar es Salaam   Cape Town
Melbourne   Auckland   Madrid
and associated companies in
Berlin   Ibadan

Oxford is a trade mark of Oxford University Press

Published in the United States
by Oxford University Press Inc., New York

British Library Cataloguing in Publication Data
Data available

Library of Congress Cataloging in Publication Data
Anand, Paul.
Foundations of rational choice under risk/Paul Anand.
p.   cm.
Includes bibliographical references.
1. Economics, Mathematical.   2. Economics—Methodology.
3. Rational expectations (Economic theory)   I. Title.
HB135.A59 1993   330'.01'51—dc20   93–17536
ISBN 0–19–823303–5

1 3 5 7 9 10 8 6 4 2

Typeset by Datix International
Printed in Great Britain
on acid-free paper by
Bookcraft (Bath) Ltd.
Midsomer Norton, Avon

In memoriam matris,
and for my family

Preface and Acknowledgements

Were you to line up all the economists in the world, they would not, it is said, reach a conclusion. Up to the start of the 1980s, the exception that proved the rule was the widespread agreement that agents were rational and that subjective expected utility (SEU) was what they maximized.

Utility has a long history in economics, but there were no foundations so sure or popular as the axioms devised by John von Neumann and developed in collaboration with Oscar Morgenstern. The von Neumann–Morgenstern version of the theory allowed economists to bring the powerful probability calculus to bear on agents' beliefs, to provide behavioural foundations for the concept of utility, to categorize and analyse attitudes to risk, and all in a mathematically tractable framework. After von Neumann and Morgenstern, utility theory had an axiomatic foundation which promised the exactness, transparency, and sheer elegance of Euclidean geometry. As the song says, who could ask for anything more?

In the 1940s axiomatic theorizing represented a quantum leap in our thinking about how mathematics could be used to model empirical phenomena in the social realm. However, by the early 1950s storm-clouds were already beginning to form in the shape of experimental evidence, some produced by the French Nobel Laureate economist Maurice Allais, indicating that the axioms were not, in fact, supportable.

To the defence of SEU came the methodologically sophisticated mathematical statistician Leonard J. Savage, who proposed an alternative interpretation. Assumptions of SEU should be read, he claimed, as canons of rationality, and as such they would be immune from empirical falsification. This was a natural suggestion for a statistician to make: it was both a model of SEU's axioms and a philosophical claim about the nature of rationality, and it was one that many economic theorists were happy to accept.

This 'interpretation hypothesis' successfully held off the cavalry for a while only. During the 1970s, the deluge of experimental falsification began to spill over into the economics literature, and the publication of Daniel Kahneman and Amos Tversky's prospect theory in *Econometrica* and of David Grether and Charles Plott's work on preference reversal in *American Economic Review*, both in 1979, marked a turning point. Utility theorists had always known of the existence of such experimental findings, but now, at last, they were out in the open.

The response from theorists was swift, and by the mid-1980s there were many new competing theories that explained what SEU could not. It was, indeed, an exciting time. Of course, if Savage was right, these new theories were only theories of *irrational* choice.

The rational status of SEU violations is an issue that has interested me for the past decade. In 1982 I published an article in which I argued that the rational agent might be able to violate at least three axioms of SEU. Since then I have sought to test that claim against the arguments of others. While my aim has been to better understand the status of normative claims common in economic theory, this is a question I took to be a matter of philosophical theory—and still do.

I write, in the first instance, hoping to interest other students of decision theory, whatever their original discipline. Those who work in the area will be used to the mix of experimentation, mathematics, and philosophy on which the subject is based. The 1980s witnessed a flurry of seminar and publication activity on the interface between economic theory and analytical and moral philosophy, and I believe that this essay will be of use to participants in these debates. This is not of itself a technical project, and will be easily accessible to final-year undergraduates interested in the foundations of economics, whether in economics of philosophy.

The first three chapters set out the subject matter and motivation for the investigation. Chapter 1 outlines key developments in SEU, emphasizing generalization, both technical and conceptual, as a driver of change in decision theory. Chapter 2 reviews some of the experimental evidence concerning rational choice.

The positive status of SEU is evaluated with the help of experi-mental evidence relating to its axioms. In addition, work on failure in logical inference is also discussed, for it both illustrates the extent of human 'irrationality' (if only in a narrow laboratory sense) and shows that even such behaviour is amenable to theoretical analysis. Chapter 3 discusses the range of objectives that decision theorists pursue and identifies the normative sense in which SEU must, ultimately, be judged.

The three chapters that follow analyse the normative status of SEU's main axioms. Chapter 4 identifies what I believe to be the key arguments for transitive preference and considers their logic. Instances of transitivity violation and their theoretical descrip-tions are also discussed. Chapter 5 examines independence, and Chapter 6 considers completeness and continuity. Both chapters review arguments raised by myself and others, and particular attention is given to interpretational issues in the discussion of completeness.

The last three chapters deal with methodological and founda-tional issues. Chapter 7 examines a proposed line of defence for SEU which saves the theory only at the expense of behavioural testability. The Socratic role that logical arguments play in establishing normative appeal is also discussed and is shown not to be as self-evident as might be supposed. Finally, in Chapter 8 I sketch an answer to the question that rejecting SEU as a theory of rationality opens up: What, if not SEU, does rational choosing entail? Chapter 9 summarizes and identifies some future directions for research.

While I try to deal even-handedly with SEU, it is only proper that I declare straightaway that in my view neither its rational model nor its positive model is tenable. My claim is just that SEU is not general enough to be the theory of choice for all rational agents. The argument is progressive for decision theory and it is one that a growing number of decision theorists are coming to hold. I shall not be advocating that we dispense altogether with SEU, though I do suggest that it is used—when it is—for reasons that are essentially pragmatic.

Chapters 7 and 8 are slightly modified versions of articles that appear in *Annals of Operations Research* and *Oxford Economic Papers*. Material in Chapters 4, 5, and 6 has been taken from work in *Theory and Decision*, the *Economic Journal*, and *Oxford Agrarian Studies*. Discussions of my own experiments in Chapter 2 are based on reports in the *Greek Economic Review* and *Agricultural Economics*. I am grateful to the editors for allowing me to use material from those journals. Many were kind enough to comment on those papers, including: Randall Bausor, Peter Bowbrick, Wilhelm Brandes, John Broome, Peter Earl, Bill Gerrard, Alan Hamlin, David Kelsey, Tony Lawson, Duncan Luce, Mark Machina, Gay Meeks, Philip Petit, Carol Propper, Gavin Reid, Jochun Runde, Howard Sobel, and Arnis Vilks. For reading earlier drafts, I am grateful to Benn Steil and especially to Bob Sugden for many penetrating and constructive insights.

In addition, some of the material has been presented to a number of audiences. I should like to thank, for their comments, seminar participants in Cambridge, Edinburgh, Oxford, and York as well as attendees of the *Second Conference on the Foundations of Utility and Risk* in Venice, 1984. In a form recognizable as the one it now has, Chapter 4 was presented to the *Recent Advances in the Philosophy of Science* conference in Amsterdam, 1991, and to the Royal Economics Society conference in London, 1992.

I am indebted to the Royal Economics Society for the Post-doctoral Fellowship that enabled me to do much of the work discussed here. Over a period of almost three years I was favoured with a number of affiliations, and I thank especially Tony Culyer, Keith Hartley and the directors of the Centre for Experimental Economics, John Hey, and Graham Loomes, all at the University of York; the Department of Applied Economics and Darwin College Cambridge, where I held a research associateship; and finally Steve Nickell, for the hospitality extended to me by the Institute of Economics and Statistics, Oxford for two terms. Experiments mentioned in Chapter 2 were funded by the Nuffield Foundation.

Although this essay is in no way an attempt to 'write up' my D.Phil., it draws heavily on ideas that I began to explore there.

My doctoral supervisors, George Peters and G. T. Jones, were both extremely generous with their time and encouragement. For this I owe them much. Others who made a number of useful suggestions at that time include my examiners Paul Webster and Donald MacLaren as well as Paul Seabright, Alan Manning, and particularly Robin Cubitt, who commented patiently on far more material than I had any right to expect. I am especially grateful to my college, Wolfson, for providing such a stimulating environment in which to do research, and to the Milk Marketing Board for the graduate scholarship that made it possible.

Last, but certainly not least, I acknowledge, first, the efforts of Andrew Schuller, who has displayed, variously and in abundance, the qualities of Mercury and Job that seem to be the prerequisites of an OUP editor. Secondly, I thank Tessa M. Shaw, who has put up with the chores and histrionics that ensue from being partnered to the unbearable person-who-is-writing-a-book with characteristic grace and skill. My inability to finish this essay sooner I blame, in part, on our sons, Conor and Donal, but I can think of no happier excuse. I thank also the President and Fellows of Templeton College for the generous opportunity my current fellowship afforded me to complete the project. My gratitude is poor recognition of the fact that this enterprise would have been impossible except for the intellectual inspiration and support of Michael Bacharach and Amartya Sen.

PAUL ANAND

Oxford
May 1992

Contents

I

Utility Theory

INTRODUCTION

Decision theory is about choosing the act that is best with
respect to the beliefs and desires that an agent holds. Subjective
expected utility (SEU) tells us how to do this, and, although
there are numerous rival theories, it is to SEU that classical
decision theory owes its structure and many of its results. Like
its competitors, SEU is a mathematical theory of decision-
making with which are associated procedures for measuring
desires and beliefs. It proposes a rule for their amalgamation
and a set of concepts that serves to interpret and justify its
formal elements.

Though its roots can be traced to early developments in
mathematical statistics, the social science applications of SEU
have been most thoroughly explored by economists. More
recently, philosophers, psychologists, management and political
scientists, biologists, and legal theorists have turned to decision
theory for the modelling tools it provides. Given the prudential
nature of social interaction, it is not surprising that the discipline
is becoming a lingua franca for researchers in a number of
disciplines, although much of the subject's success turns on the
appeal of just one theory, SEU. This essay considers an issue
that arises predominantly on the interface between economic
theory and philosophy but is of relevance to anyone interested
in the foundations (or uses) of decision theory. It is often
claimed that the axioms of SEU identify canons by which
rational agents must be guided, and it is this view that I want to
suggest we have no grounds to accept.

I shall say more about the background to this position in
Chapters 2 and 3, but it may be helpful to say something

immediately. Those accustomed to, and indeed attracted by, the mix of experimentation, mathematics, and philosophy that characterizes the practice of modern decision theory will recognize this for the normative, theoretical project that it is. Some economists, however, mindful of the claim that, via utility maximization, assumptions like transitivity underpin *neo*-classical economics, will want to know about the implications for practice. It is important, therefore, that we begin by acknowledging the mathematical economic results which show how theories of general equilibrium, consumer behaviour, and utility maximization (to name but three) can be generalized far beyond the boundaries of SEU. The view that rationality is defined by SEU commits one to saying that these generalizations are theories of irrational action, a position that for many would be at least as unwelcome as dropping the assumption of rationality.

Mark Machina (1989) proposes a compromise. He is willing to concede that SEU is normatively desirable at some suitably deep level of primitive description. In return, he suggests, defenders of SEU should accept that economists rarely have access to those deep-level descriptions. Machina's express aim is to encourage the use of non-SEU theories. Indeed, the substance of his review makes a very elegant and important philosophical contribution to the foundations of decision theory, although I find the final conclusion, we might call it the 'M twist', surprisingly open-ended.

In any case, my aim is to seek a resolution to the rationality question *per se*. A corollary shared with Machina's argument is that generalizations of SEU are not, by virtue of their weaker structures, theories of irrational choice. However, while I have no wish to become embroiled in a 'more-or-less-mathematics?' debate, I shall suggest that a subtler reading of the use of mathematics in social science is warranted[1]. In this chapter I introduce SEU by discussing some of the problems that it solves, its axiomatic basis, and a few of the many issues to which it has, successfully, been applied.

---

[1] Work of related interest can be found in Lewis (1988) and Vilks (1992).

THE ORIGINS OF DECISION THEORY: FROM EXPECTED VALUE
TO EXPECTED UTILITY

In many respects, decision theory reflects, at a micro level, the liberal values of a modern democracy. It does not seek to prescribe what agents should want (or believe), and in this its origins can be traced back not, as one might expect, to Mill's account of utility,[1] but to David Hume's remark that reason is the slave of the passions. This instrumental view of rationality holds that reason is just a tool by which actions, that lead to satisfaction, are chosen.

A second source on which the subject draws can be found in the history of statistics and owes much to the analysis of betting games (seemingly conducted, for reasons best left to speculation, by members of the ecclesiastical profession). The availability of demographic statistics and the vicissitudes of maritime trade also played a role in driving the demand and supply of systematic methods for evaluating uncertain prospects; and as early as 1657,[2] there appears to be a discussion of a rule still used in statistics today. The rule is what we call 'expected value', and it requires that the worth of an action be estimated as the sum of pay-offs weighted by their probabilities.

In common parlance, 'pay-off' is well understood, although it can refer to either a monetary value or the subjective worth a person attaches to an outcome. The first meaning is used by the expected-value rule and, though intuitively reasonable, it leads to unacceptable results in a class of problems first described by Daniel Bernoulli (1739).[3]

Bernoulli's illustration of these problems, also known as the 'St Petersburg paradox', runs as follows. For a price, Peter offers Paul the opportunity to take part in a game. A coin will be tossed, and if it lands on heads Peter pays Paul 2 ducats. If

[1] Mill (1863) proposes a hierarchical (we would say lexicographic) theory of preference in which higher-order needs (self-actualization) are satisfied only after basic biological needs have been met.

[2] See e.g. 'De Rationciniis in Ludo Aleae' by Christianus Huygens.

[3] It seems that Cramer and Huygens may have had similar ideas at around the same time or before.

the coin lands on tails they toss again, only this time Paul will be paid 4 ducats if he wins. Every time Paul loses Peter doubles the stake, and they continue to toss until Paul wins. The question is: At what price, if any, should Peter not pay to play?

If one uses the expected-value rule, it turns out that Paul should pay any price. At the first toss, the probability of winning 2 ducats is $\frac{1}{2}$. On the second toss the prize is 4 ducats while the probability of winning, i.e. the chance of losing on the first round and winning on the second, is $\frac{1}{4}$. By repeating this reasoning, the expected value of the game can be taken as the limit of the following sum:

$$(\tfrac{1}{2} \times 2) + (\tfrac{1}{4} \times 4) + (\tfrac{1}{8} \times 8) + \ldots$$

The sum has no finite limit so the expected value of the game is infinite. To play the game, Paul should be willing to pay any price. While it is clear that following the expected-value rule gives the game an infinite worth, it is far from obvious that anyone is compelled by rationality alone to give up all his wealth to take part in a game that might end on the first round with a pay-off of 2 ducats. Indeed, it seems reasonable to say that, in so far as the opposite is true, the St Petersburg paradox serves to show the absurdity of following the expected-value rule. Bernoulli's argument is, in effect, a *reductio ad absurdum*.

The resolution of the paradox, attributed to both Bernoulli and a contemporary, Nicolas Cramer, turns on the evaluation of uncertain prospects using a non-linear function of monetary pay-offs. Specifically, it was suggested that, as a person's wealth increases, increments in income are less valuable, so that avoiding a loss of x is more desirable, say, than receiving a gain of x. Such preferences can be represented by a function which maps wealth on to utility (a real-number measure of desirability), and it was shown that if this function was logarithmic then the expected utility of the game would be finite. If the agent chooses the act that maximizes expected utility, he or she might pay only a moderate sum to enter the game—which accords better with what we think it is intuitively reasonable to do.

There are those (some natural scientists, for instance) who question the role of subjective elements, like utility, in the scientific analysis of behaviour. However, in contemporary economic thinking it goes without saying that the evaluation of risky prospects must allow for psychological aspects of worth. The importance of the St Petersburg paradox is not just that it provides a useful generalization of the expected-value rule (important as this is), but that it also demonstrates a methodological point about the compatibility of mathematical theory and respect for subjective evaluations.

SUBJECTIVE PROBABILITIES

A major concern of decision theory is how best uncertainties should be described.[4] Until this century, most researchers assumed that the relative-frequency account of probability provided the appropriate medium for addressing such issues. As a mathematical measure, the conventions governing the use of probabilities are reasonably uncontroversial; for instance, the probability that two mutually exclusive events will occur simultaneously is zero, while the probability that one of all the mutually exclusive events will occur is unity. Less agreement, however, exists about what probabilities are (the metaphysical question) and how they should be measured (the epistemological question).

A number of theories suggest how these questions might be answered, and they do this by providing a model, that is an interpretation, of the abstract probability calculus. The classical approach, associated particularly with Laplace (1951), treats probability as a measure of ignorance about deterministic events and suggests that, in the absence of sufficient reason to think otherwise, events should be taken as equally probable. Though there is some appeal in this use of the insufficient-reason principle, ambiguity surrounding what counts as an event has led to rejection of the view. If throwing a 6 and an odd number are both events, then what reason is there to suppose that they are not equally likely? However, if these are equally probable, then

---

[4] Uncertainty is nearly always a cost for producers, but this is not always the case for consumers. A predictable thriller may not make for a good film.

how can the chance of throwing a 6 be the same as the chance of throwing a 5? A second view, known as the logical or 'a priori view', suggests that probabilities are to be found in the logical relations between propositions and the availability of relevant evidence. This has the advantage of providing a definition for all possible propositions and any evidence that might be brought to bear on them. On the other hand, it leaves undetermined the precise numerical value of the probability in many situations, and it is perhaps for this reason that the enthusiasm of probability theorists for the view has not been shared by the users of statistical methods. A third view, the one most widely invoked, holds that a probability is the relative frequency of an event within a set or trials. Following Kolmogorov (1951), it is common for theorists to think of the value as being defined by the limit of a ratio of occurrences to trials as the number of trials tends to infinity. While these limits may be calculable, they are not directly observable by mortal observers, and in practice we often take the probability to be the relative frequency based on trials observed to date. Although each model contributes something to our understanding of probability, it is relative frequency that underpins the everyday business of data-gathering and statistical analysis.

Useful though relative frequency is, it is not without shortcomings. First, if we define probability as the limit of an infinite sequence, there is no reason to suppose that the relative frequency we infer from a finite subset of the infinite sequence is in any way related to the 'true' probability as defined. Second, many important economic events (e.g. the introduction of a single currency into the European Community) are unique and their likelihood is not easily thought of as the limit of an infinite sequence of similar events. Third, even for potentially repeatable events, there may be no trials on which a relative frequency can be estimated. Finally, there is a question about what the probability is in situations where different decision-makers have observed different relative frequencies.

Another, essentially decision-theoretic, interpretation of probability that addresses at least some of the problems inherent in the relative-frequency view holds that probabilities measure beliefs. The model propels probability theory into the realm of the subjective just as objective pay-offs are treated according to their psychological significance via the concept of utility. As decision-makers can have beliefs about future or unique events, subjective probabilities may exist in situations where relative frequencies do not or could not. In addition, the subjectivist is not embarrassed by the possibility that the measured probability of an event is not unique. Different decision-makers have different information sets, so there is no reason to suppose that the beliefs, even of rational decision-makers, are the same.

If there is a weakness with the subjective view, it arises from the measurement process. To determine a person's beliefs we need to know the contents of other minds, and it is far from obvious how such access is to be obtained. It would, therefore, be of considerable interest if we could show that beliefs *can* be measured, and this is precisely the contribution that the philosopher–mathematician Frank P. Ramsey (1931) made. Most applied uses of subjective probability estimate beliefs from answers to direct questions. The instrument that Ramsey proposed is so different that the use of the term 'subjective' to cover both meanings is misleading. In contrast to asking people directly for the probability that it will rain tomorrow ('pick a number between zero and one'), Ramsey showed that, if decision-makers could be persuaded to say how they would *bet* on outcomes, the subjective probabilities could be inferred from their responses. It is the Ramsey sense of subjective probability to which discussions of the foundations of decision theory refer: it is a belief revelation theory, analogous to Paul Samuelson's work on revealed preference, and it is the one we shall now examine.[5]

We note that Ramsey's is a procedure for eliciting beliefs based on choices over lotteries.[6] For this purpose, we think of a

---

[5] Where decision theory is being applied to help people come to a decision, directly elicited subjective probabilities are used to the complete exclusion of inferred beliefs. There is a substantial, prescriptive literature for probability elicitation, and many of the key points are collated in Spetzler and Stael von Holstein (1975).

[6] This version draws, in part, on an excellent exposition due to Jeffrey (1965).

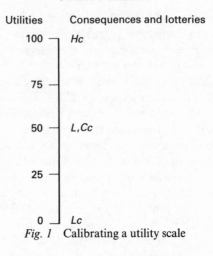

Fig. 1   Calibrating a utility scale

lottery as a set of 'if–then' propositions[7] and describe the ith lottery, $\mathscr{L}_i$, as {if Xs then Ac, if Ys then Bc, etc.}, where Xs is the state X, and Ac the consequence A. To measure beliefs it is necessary to allow for differences in preferences, which we do by developing a scale of utilities. From a measure-theoretic viewpoint, utility is like time, so in the first instance two distinct consequences can be assigned to any two numbers. Equivalently, it is possible to choose an origin and a unit of utility that happens to be convenient; in applications, if it is known that Hc and Lc are the best and worst consequences to be scaled, then one convention assigns to Hc, the highest consequence, a utility score of 100, and to $u(Lc)$, the utility of the lowest consequence, a score of 0 (see Fig. 1).

Next we find a state, Es, that has an evens chance of occurring: $p(Es) = 0.5$. In addition, Es should have no bearing on the utilities of Hc and Lc. Using this *ethically neutral* state, as Ramsey called it, the following lottery, L, is constructed: {if Es then Hc; if $\neg Es$ then Lc}.[8] Weighing the utilities of Hc and Lc by

---

[7] Whether these are conditionals or counterfactuals is not of consequence except for a discussion of transitivity in Ch. 4.

[8] $\neg$ is not, the logical connective.

the probability of Es and $\neg Es$, the lottery is assigned a utility of $0.5(100) + 0.5(0)$. The lottery's utility score of 50 gives, then, the third, mid-point of the scale.

At this stage, the agent whose beliefs are being elicited begins to play a more substantive role. He or she is asked to consider whether there is a sure pay-off which would be just as desirable as the $50:50$ lottery on the extremal outcomes. This amount, call it Cc, known as the certainty equivalent of the lottery, is assigned a utility of 50 because it is as desirable as the lottery with that utility value. This certainty equivalent can be used to construct two further lotteries: {if Es then Hc; if $\neg Es$ then Cc} and {if Es then Lc; if $\neg Es$ then Cc}. As before, the utilities of the consequences are weighed by their probabilities to derive utilities for these lotteries. The utility of {if Es then Hc; if $\neg Es$ then Cc} is $0.5(100) + 0.5(50) = 75$; similarly, the utility of {if Es then Lc; if $\neg Es$ then Cc} is 25. By repeating the process, as many points on the scale can be defined as materiality demands. Once the utility scale is complete, it is possible to assign a utility value to any lottery (or state) simply by comparing it against the lotteries whose utilities are already known.

Once any lottery can be given a utility, subjective probabilities for events can be derived. Suppose it is necessary to ascertain a person's beliefs about the likelihood of some state Us. First, we obtain the utilities of the lottery $\mathcal{V} = $ {if Us then Xc; if $\neg Us$ then Yc} and its component states Xc and Yc—call these u_3, u_1, and u_2 respectively. If p denotes the belief that Us will occur, the utility of $\mathcal{V}$ is given thus:

$$u_3 = pu_1 + (1 - p)u_2.$$

Rearranging terms, the following expression for p can be derived:

$$p = (u_3 - u_2)/(u_1 - u_2).$$

Surprisingly, a subjective probability can be thought of as the ratio of two utility differences. Ramsey shows that direct questions about probabilities that are not always easy to answer can be avoided. His procedure prompts us to seek evidence for beliefs not in what people say, but in what they do, and it does so in a way that is as objective as any offered by competing theories of probability. It does not preclude the possibility that decision-makers base their beliefs on observed relative frequencies, but it does obviate the need to conclude that decision-makers have no probabilities where, for whatever reason, relative frequencies are unavailable. In purely aesthetic terms, it is difficult not to be awestruck by the simplicity and efficiency with which Ramsey establishes one of the key results in decision theory.

Like other models of probability, the subjective view has its drawbacks. In particular, one might be concerned that these subjective probabilities seem to be based solely on choices that are arbitrary (except for the individual concerned). If the choices *are* arbitrary, then so, presumably, are the beliefs. As it happens, it can be shown that the choices one makes must adhere to certain conditions if utilities and probabilities are to be derivable. These conditions are the axioms of SEU. Because only certain choice patterns are consistent with SEU, it is reasonable to say that, in at least one sense, elicited beliefs cannot be arbitrary. As we shall see in the next section, it is precisely this sense of non-arbitrariness that has been taken as the formal theory of rational action.

THE SAVAGE AXIOMS

Axiomatic methods have curried considerable favour with mathematical economists. In so far as the approach uses logic embedded in mathematical terminology to derive consequences from simple, clearly stated premises, this can hardly be thought of as a radically new way of constructing theory. What distinguishes these theories is the particular meaning of axiom that is presupposed. Axioms are intended to be not just postulates but rather assumptions which no reasonable person could reject. Proponents of SEU have emphasized this reading, and in so doing they have raised the importance of axioms in the task of theory evaluation. To the extent that we focus on axioms, we allow, as we must, SEU to play on its home ground.

There are many axiomatic versions of expected utility theory—see Fishburn (1981) for a review. The von Neumann–Morgenstern (1944) account takes probability as given and offers a method for eliciting utility functions not dissimilar to that devised by Ramsey. It was this version that popularized among economists the modern interpretation of utility as sensitivity to risk. The influence of this version is reflected in the widespread use of the term 'von Neumann–Morgenstern utility function' and it is, undoubtedly, one of the milestones in the application of mathematical methods within economics.

It is, however, to Savage's account (Savage 1954) that statisticians and mathematical economists usually refer, and the difference is not entirely conventional. In the first place, unlike Ramsey or von Neumann and Morgenstern, only Savage provides axioms with which measures of inferred subjective probability and utility are consistent. Secondly, the distinction between positive and prescriptive uses of SEU did not come clearly into focus until the version due to Savage. Indeed, the first half of his book is still the most substantial attempt to find a single, defensible interpretation of SEU.

Although SEU has been, and continues to be, used as a description of how decision-makers behave, it is worth stressing that Savage's version was not intended to be a contribution to empirical science. Rather, he proposed a model that is uncompromisingly philosophical. As an account of what the rational agent must do, it has behavioural content, but it is not behaviourally testable. The postulates of SEU are ideals like the axioms of geometry, although they differ in one rather fundamental respect. Plato's account of the relationship between geometric idealizations and real objects in the physical world does not lead one to suppose that any particular line will have the zero width which geometry makes it useful to assume that lines have. Savage, on the other hand, does expect that real decision-makers who violate the assumptions of SEU should modify their actions so that they do, literally, conform.[9]

___

[9] Methodologists might note that this runs counter to ideas like those of Friedman (1953), who plays down the importance of assumptions. His ideas are

With this background in mind, we may examine the axiomatization of SEU, beginning with the definition of a number of primitive terms. The *world*, Savage says, is any object about which the person is concerned; a *state* (of the world) is a description leaving no relevant aspect undescribed; and *the state of the world* is that description which is the true description (Savage 1954: 9). In addition, he defines a *consequence* as anything that may happen to you and notes that the existence of a decision problem presupposes the need to choose between two or more acts. Acts themselves are completely identified by their possible consequences, so that '[i]f two different acts had the same consequences in every state of the world, there would from the present point of view be no point in considering them two different acts at all.' He also notes (p. 17) that '[l]oosely speaking, [the person *prefers* an act f to an act g] ... if, were he to decide between f and g, no other acts being available, he would decide on f'.

These primitives are sufficient to understand Savage's axioms described below. In later chapters we shall discuss the normative cases for and against completeness, transitivity, (sure-thing) independence, and continuity in probability. If you have not encountered these axioms before, this is an opportunity for you to form your own opinion about their normative status.[10]

Axiom 1: Completeness

For all acts x and y, either x is preferred to y or y is preferred to x or x and y are equally desirable. Note that the 'greater than or equal to' relation, $\geq$, defined over real numbers, has the same property. Only three situations are possible: either m is greater than n ($m \geq n$ and not $n \geq m$), or n is not greater than m

---

much discussed in general economics methodology but have not been used by economists proposing mathematical axiomatizations of utility theory, nor do they carry much weight in decision theory itself. It is dangerous, therefore, to discuss the methodology of utility theory without first specifying the context in which it is being applied.

[10] Where matrices are used to illustrate the force of an axiom, acts and states are listed on the vertical and horizontal axes respectively and consequences are entered in the matrix cells.

($n \geqslant m$ and not $m \geqslant n$), or m and n are equal ($m \geqslant n$ and $n \geqslant m$). It is beyond question that size-based comparisons between real numbers have the completeness property, so if comparing measures of preference is analogous to comparing the sizes of numbers, then preferences will also be complete.

Axiom 2: Transitivity

For all acts x, y, and z, if x is preferred to y and y is preferred to z then x should be preferred to z. Again, the analogy with comparisons between real numbers is relevant. We know that the $\geqslant$ relation is transitive for all real numbers and that if the value of an option is measurable then comparison of the measures of desirability will necessarily be transitive. This is simply a matter of logic. However, suppose that, for three commodities, a decision-maker claimed to prefer a to b, b to c, and yet prefer c to a, that is, had *in*transitive preferences. Such a person would, it seems, be willing to pay to trade a for c, c for b, and finally b for a. If repeated, such behaviour could leave a person destitute for no apparent gain. This attempt to provide a *reductio ad absurdum* is not the only way of grounding transitivity, but it is one of the most commonly used methods and one to which we shall return in Chapter 4.

Axiom 3: The Independence Axiom

The third axiom is more easily explained with the help of two decision problems, the details of which appear in the act–state matrix of Table 1.1.

Table 1.1

	Event A	Event $\neg A$
Decision 1		
Act w	k	m
Act x	k	n
Decision 2		
Act y	l	m
Act z	l	n

Suppose there are two events, A *and* $\neg A$, and that the consequence of each event depends on the act chosen. Independence requires that, if w is preferred to x (or vice versa), then for all consequences k and l, y is also preferred to z (or vice versa). As with other independence axioms, the justification depends on an application of the principle of insufficient reason. In decision problem 1, the only consequences that distinguish the acts are m and n and these are, therefore, the only consequences that could give rise to a reason for strict preference. However, they are also the only consequences that distinguish acts in decision problem 2, and because they make exactly the same distinction between y and z as between w and x, the reason for choosing which they provide is invariant between the decision problems. If they provide a reason for a strict preference at all, so the story goes, then they provide a reason for the 'same' preference in both problems.

Axiom 4: Resolution Independence

If, *ex ante*, two acts yield consequences k and l, then the preference for those acts after an event depends only on the preference between realized consequences—not between ones that might have been.

Axiom 5: Expected Wealth Independence

If a decision-maker is about to place a stake on one of two lotteries, the decision should be based on the probability of

Table 1.2

	Lottery 1		Lottery 2	
	Event A	Event $\neg A$	*Event B*	Event $\neg B$
Act w	k	$k - \varepsilon$		
Act x	m	$m - \delta$		
Act y			k	$k - \varepsilon$
Act z			m	$m - \delta$

[11] An 'event' in Savage's framework is a set of states.

winning but not on the amount of the stake. Table 1.2 provides an illustration of this idea. Assuming that δ and ε are strictly positive, if an act w is preferred to act y (or vice versa) then act x is preferred to act z (or vice versa). In other words, if A is believed to be more probable than B, then the preference for lottery 1 should not depend on whether prize k or m is greater.

Axiom 6: Minimal Strict Preference

There is at least one pair of consequences such that one is strictly preferred to the other.

Axiom 7: Continuity in Probability

There exists a state with a strictly positive probability so small that any change to the pay-off in that state, however large, will not affect the preference for that act. This is, in effect, a decision-theoretic parallel to the statistical convention that very unlikely events should be regarded as having a zero probability.

Axiom 8: Partial Resolution Independence

If an act x is preferred to each and every consequence of another act y for a set of states, then x is preferred to y if the event comprising the set of states obtains.

Using axioms 1–6 (A1–A6), Savage shows that a probability ranking can be inferred from a decision-maker's behaviour (see Savage 1954 or Fishburn 1988*a* for proofs). Of these, A1–A3 have received most attention in the literature, and it is those axioms that will be the main focus of discussion in subsequent chapters. Of the remaining premises, little objection would be made to A6 if it is read as limiting application of theory to situations where the decision-maker knows enough about him- or herself and the meaning of the consequences to have a strict preference. A4 was designed to 'assert that knowledge of an event cannot establish a new preference among consequences' (Savage 1954: 26), while A5 summarizes the idea that the alternative you bet on 'does not depend on the prize itself' (p. 31) but only on its probability. These are usually regarded as having no substantial interpretation and I shall not consider them, though the reader may notice that some of the issues they raise are similar to those arising from the independence axiom, A3.

In addition to providing an axiomatic sub-structure for SEU, Savage makes a cognitive claim that bears on the motivation for using inferred subjective probability. To be precise, he suggests that it is difficult for human agents to answer questions about probabilities though they have no such difficulty in answering questions about choice between gambles. Savage is right, I think, to imply that we need some reason for not being content with directly elicited subjective probabilities, although this particular reason is not without its problems. Even if it were true, the point could be made that ease of reply might indicate that the true, difficult nature of the task has been obscured. Empirically, decision analysts have found that, with the aid of procedures that support the elicitation process, direct questioning seems to work well. Furthermore, evidence based on work with US weather forecasters, *inter alia*, shows that frequent predictions with feedback can lead to expressed subjective probabilities that are very accurate when compared with the appropriate relative frequencies.

Some might argue that the weakness of the cognitive motivation is a side-issue, as the use to which SEU is put by neoclassical economists, for example, depends merely on the constructability of a subjective probability distribution. In any case, analytical results that presume only the existence of such a distribution do not, after all, depend on particular probabilities being feasibly recoverable. The other main source of interest in the SEU axioms derived from Savage's hermeneutic hypothesis, which is that the assumptions of SEU can be interpreted as canons of rationality.

Ramsey, in contrast, takes a more approximate attitude to the theory and explains his avoidance of axiomatization thus:

I have not worked out the mathematical logic of this in detail, because this would, I think, be rather like working out to seven places of decimals a result only valid to two. (Ramsey 1990: 76)

My reading of the sense in which SEU should be used is much closer to this view than that which came to prominence within classical decision theory. Savage's interpretation hypothesis is unnecessary according to Ramsey, and it is that hypothesis that this essay seeks to test.

THE USE OF UTILITY

We said that subjective expected utility is the culmination of research into decision-making over a period of almost three hundred years. Key to its development and acceptance have been the introduction of mathematical elements and the construction of new interpretations of those theories. The generalization of expected value to expected utility as a means for evaluating gambles, the interpretation of Bayes's theorem[12] as an account of belief-updating, Ramsey's behavioural method for estimating utilities and probabilities, and Savage's axiomatic foundations are all contributions to mathematical statistics, although the greater influence has been on the conduct of economic analysis. Each of these contributions pushes decision theory away from the idea that uncertain prospects can be objectively evaluated.

It is an essential feature of SEU that it specifies a mathematics of decision, but it is open whether we use it for normative or positive purposes. A philosopher of science might want to say that at its core it is not theory but a partially interpreted model from which theories are constructed. That seems to me a flattering but fair light in which to see the contribution to decision theory that SEU makes. In decision theory, it provides a basis for using the maximization of subjective expected utility to select actions where outcomes are not certain. The contribution to conventional game theory is twofold. In the first place, and following von Neumann and Morgenstern (1944), we take the consequences of game theory to be utilities for which SEU

---

[12] This is not an axiomatic element of SEU, but it is associated with the subjectivist doctrine and will be mentioned in Ch. 8. The interpretation to which I refer is of the theorem for relating conditional probabilities as a method for combining information.

provides the means of measurement. However, it is also true that the majority of games have no game-theoretic solution. Players in them can optimize only if they provide subjective probabilities for opponents' strategies, so most games are in fact decision problems. In economics, SEU is used as an instrument to provide a description of the reasoning agent which is embedded in theories of group behaviour, such as consumer theory and general equilibrium. In experimental research conducted by psychologists and economists, the axioms of SEU have been used to make behavioural predictions which are then tested under laboratory conditions. In decision analysis and management science, SEU is used to help decision-makers by providing a structure in which objectives and trade-offs can be made explicit. It is hard, as I have said, to overestimate just how useful SEU has been and how valuable no doubt it will continue to be to users of decision theory for a long time to come.

This essay is concerned primarily with the normative justification of SEU, notwithstanding the traditional view of science as a positive enterprise. Normative goals that scientists (decision theorists and economists particularly) pursue, on the other hand, are rarely discussed in the philosophy of science literature, although they are essential to many of the projects that we set ourselves. In a sense, empirical violations of SEU provide a *raison d'être* for the normative interpretation, and it is these to which we turn in the next chapter.

2

Experimental Evidence for
Rational Choice

INTRODUCTION

As a scientific theory about human decision-making, SEU is false, and it is false in two senses. First, there are situations in which agents choose counter to the axioms of SEU, a result that constitutes falsification in the Popperian sense. In many experiments, the choices that run counter to SEU actually outnumber the occasions in which choices are consistent with the theory. This means that SEU cannot be regarded as even approximately true, and that its use would lead to worse predictions than could be made on the basis of random guessing. Second, it is sometimes said that, because random elements are bound to enter into human deliberation, only systematic errors should be of concern. Even by this criterion, the preponderance of certain types of violation suggests that at least some are systematic and that SEU is, therefore, biased. A large body of empirical evidence has been reviewed both by psychologists and by economists (Rapoport and Wallsten 1972, Schoemaker 1982, Machina 1983, and Baron 1988), and all come to similar conclusions: if SEU is an empirical, testable theory, then it is, in any conventional sense, untrue.

This chapter seeks to review some of the experimental evidence relating to SEU. If the normative interpretation is correct, then it must be accepted that human beings are often not rational, even when offered the cognitive luxury of simple laboratory tasks. Some researchers have proposed a modest response. If people persistently choose in particular ways, then perhaps it is we who should question our account of rationality. There is something in this, though there are limits to which such a view can be taken. For instance, there are very few decision theorists or economists who say that 'one-boxing' in Newcomb's problem is reasonable, however many subjects do it.[1] The equation of rationality with consistency seems to presume that logical consistency is part of the rational agent's make-up. We begin by looking at experiments on logic-based choice and reasoning because if there can be any behavioural evidence of irrationality then logical inconsistency is, I believe, its safest source.[2]

LOGIC

Correct logical inference, though not overtly assumed by SEU, is a reasonable assumption to impute to models that require rationality. Certainly the assumptions that agents make non-trivial mathematical calculations, or that they behave as if they do, and that markets work to select out those agents who do not, all presuppose that agents come to views that are mathematically correct. They might do this any way they choose, but in the absence of a plausible alternative, the obvious solution is to assume that agents use tools isomorphic to those used by theorists (not consciously, perhaps), and that these are essentially mathematical and at the very least logical.

A *locus classicus* for the related experimental literature is Wason (1968), who reports an experiment in which subjects were given a task based on the following proposition:

> (P1): If there is a vowel on one side [of a playing card] then there is an even number on the other side.

Lying flat on a table before the subject were four cards, each bearing (a different) one of the following letters or vowels on each of its faces: E, D, 4, and 7. Every subject was asked which

---

[1] The problem is explained in the section of this chapter entitled 'Evidential Reasoning'.

[2] Let us admit that no source is perfectly safe. Researchers in artificial intelligence have argued that the apparent irrationality may be an artefact of the stylized-choice problems used.

cards he or she needed to turn over in order to test whether (P1) was true. Based on logical considerations alone, the rational choice is to turn over the E and 7 cards, though fewer than 10 per cent of subjects made this choice—most opted for E only.

On the face of it, the modal behaviour is normatively incorrect. The fact that the responses cluster around the E card indicates that these errors may not be arbitrary. Johnson-Laird and Wason (1970) proposed that the result was evidence of a bias towards verificationism. If the 7 card is selected, its obverse might lead to the falsification of the proposition. On the other hand, if the E card is selected and its obverse is an even number, the card-faces verify the proposition in the sense that the conjunction of E and 7 is an instance of the rule expressed in (P1). However, there would be no grounds to reject (P1). Subsequent work by Evans (1982) showed that subjects pick any cards that display instances of objects specifically mentioned in the antecedent, a phenomenon he called the 'matching' bias.

Further insight into logic-dependent choice comes from work on the 'ecological validity' of results. Wason and Shapiro (1971) found that, when the abstract cover stories are exchanged for more realistic ones, performance improves considerably. Given statements such as 'Every time I go to Manchester, I travel by train', and 'If a letter is sealed, then it has a 5d. stamp', the proportion of choices that were correct rose to 62.6 per cent; only 12.5 per cent of responses were correct in abstract settings. To explain this result, Wason and Shapiro propose an 'insight' theory which suggests that human subjects do indeed possess a mental logic (akin to the ability to do arithmetic mentally), but we arrive at incorrect responses because we fail to appreciate the exact logical nature of the task. The fault lies in the identification, rather than the solution, of problems.

Manktelow and Evans (1979) add a further dimension to this story by identifying problems in which realism has no effect on behaviour. Combining this result with those from previous studies, Griggs and Cox (1982) proposed a memory-cuing hypothesis. Realism can facilitate normatively appropriate behaviour, they argued, but only when it evokes memories of solutions to problems similar to those the subject has already experienced.

By changing the referents of the Wason–Shapiro statements to situations of which subjects had no previous experience, it is possible to provide a simple and relatively unambiguous test of the contribution that memory makes. As it happens, there is no change in the proportion of correct responses using propositions about realistic but novel situations, which suggests that, if memory does play a role, it is not the role that Griggs and Cox suggest.

More recent explanations of behaviour that depends on logical inference suggest that we analyse problems with the aid of localized schema called 'scripts' (Schank and Abelson 1977), or 'frames' (Minsky 1975). According to these accounts, the ability to reason correctly depends on the subject's ability to find a known instance of a well understood problem sufficiently analogous for that understanding to be applied in the new context. The rationality with which a person decides depends on the interaction between the nature of the problem and the level of experience. Such a view, if correct, would seem to require that theories of decision-making are tested using information that goes beyond that which can be found in the laboratory, collected perhaps by methods more usually associated with ethnography. The feasibility of the combination is demonstrated in an elegant paper by Christina Gladwin (1980), in which she uses the elimination-by-aspects model, Tversky (1972), to explain crop choice among Guatemalan farmers.

If rationality is about consistency, including logical consistency, then human agents cannot be considered rational in the full sense. People do improve as tasks become more familiar, but even then a substantial minority of choices seem to be illogical. Even if the axioms of SEU do also specify rationality constraints, we should not be surprised if these are violated. A consequence of the argument I propose is that violations of SEU should not be regarded as evidence of irrationality. People do, none the less, violate the assumptions of SEU, as we shall see in subsequent sections of this chapter.

INDEPENDENCE

Of all SEU's assumptions in Savage's axiomatization, it is the independence axiom that, until the early 1980s, attracted most attention. There are many papers reporting on experimental investigations, and, though early findings may not have been accepted, the weight of replication and improved experimental design has led decision theorists to accept that the independence axiom does not, as a matter of fact, hold. As a result, this section does not attempt to be comprehensive, but rather focuses on the kinds of independence violations that have come to light so far.

A problem due to Maurice Allais (1953)—known as the Allais paradox—provides the earliest example of a situation in which subjects consistently violate independence.[3] Variants of the problem are numerous, but in Allais's original formulation the subject is asked to consider two decision problems that differ only with respect to the pay-offs received under one state.

In Table 2.1, subjects are asked to express a preference be-

Table 2.1 The Allais paradox*

| | States | | |
	$s_1: p = 0.10$	$s_2: p = 0.89$	$s_3: p = 0.01$
Situation 1			
Lottery A	5	1	0
Lottery B	1	1	1
Situation 2			
Lottery A'	5	0	0
Lottery B'	1	0	1

* All payments are in £m.

tween lotteries A and B in situation 1 and between A' and B' in situation 2. In the original experiment, it was found that the

---

[3] Allais also provided a theory that explained the results, but this seems to be an instance, perhaps unusual in economics, when it is the fact rather than the theory that posterity remembers.

Table 2.2   The Common Ratio effect

Option	Probability of winning	Prize
Problem 1		
A	p	Dinner with Mrs Thatcher
B	q	Trip to movies with Ronald Reagan
Problem 2		
A'	p·r	Dinner with Mrs Thatcher
B'	q·r	Trip to movies with Ronald Reagan

modal constellation of choices was B and A', though independence is consistent only with choices of A and A' or B and B'. A possible rationalization is that the violations reflect the fact that subjects are indifferent between A and B and between A' and B'. However, if subjects were indifferent, why should one type of violation appear consistently more often than the other?

A second species of independence violation occurs when the probabilities of winning in one lottery are fractions of those in another. The problems illustrated in Table 2.2 help to illustrate the phenomenon. In problem 1, the agent is given a choice between lotteries A and B which offer either a prize with a certain probability, or nothing. The lotteries in problem 2 differ only with respect to the probabilities of winning, which are r times as high,[4] and an implication of independence is that preference between options A and B should be matched by preference between A' and B'. Kahneman and Tversky (1979) found, however, that, although 73 per cent of subjects preferred a 0.001 chance of $6,000 to a 0.002 chance of $3,000, when the probabilities were multiplied by 45, only 14 per cent of subjects exhibited parallel preferences for the transformed lotteries. Similar results are reported by MacCrimmon and Larson (1979) and Hagen (1979).

Another bias noted by observers arises from the finding that

[4] The following qualifications should be understood: $r > 0$ and $p·r$ and $q·v \leqslant 1$.

subjects tend to place proportionately more weight on low probabilities. An experiment by Yaari (1965) was designed explicitly to distinguish between curvature of the utility function (taken to measure preference for risk) and over-emphasis of small probabilities. With this design it was possible to show that what seemed like risk-loving behaviour could not be explained by risk preferences although it was compatible with an exaggeration of small probabilities attached to the positive outcomes of the lotteries.

Finally, it is worth mentioning what has come to be known as the utility evaluation effect. As noted in Chapter 1, Ramsey proposed a method for eliciting utility functions. A similar method, and one that also assumes independence, involves asking the decision-maker to announce certainty equivalents for one-consequence lotteries (so called because only one consequence is non-zero) with different pay-offs. For example, people might be asked how much they would accept in exchange for a 50 per cent chance of winning £100, then for a 50 per cent chance of winning £200, and so on. Because all expected utility elicitation methods (must) assign utilities to lotteries in propor-

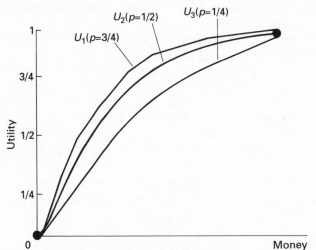

Fig. 2 The utility evaluation effect
Source: adapted from Machina (1987:131)

tion to the probability of their occurrence, the functions recovered using lotteries with a 50 per cent chance of winning should be the same as those using any other probability of winning. Experimental work reported by Karmarkar (1974) and McCord and de Neufville (1983) shows that, in practice, this is simply not the case. Fig. 2 indicates how the curvature of utility functions increases as the probability of winning rises. Increasing concavity is theoretically interpreted as higher risk aversion. The payment of lower certainty equivalents for gambles, the measure of risk aversion, is, therefore, an artefact of the probabilities used in the elicitation procedure. Whatever the cause of this phenomenon, no one utility function (strictly, no one family of utility functions) emerges as independence requires that it should.

TRANSITIVITY

The general view seems to be that in the main transitivity holds, although a few violations can be found. However, the interpretation of early experimental work on this topic seems to reflect a belief that mathematically based theory depends on transitivity.

It is useful to keep in mind the distinction between strict preference and indifference because it is generally allowed that indifference is not, empirically, a transitive relation, a view that is expressed both in philosophy, e.g. Schumm (1987), and in economics, e.g. Machina (1983). The basis for allowing intransitive indifference is cognitive and is often based on thought experiments of the following kind. Suppose that in front of you there is a line of chocolate sections arranged in ascending order of weight, and that each piece differs from its neighbour by no more than 0.01 gram. In every choice between neighbours you are indifferent, and yet when it comes to choosing between a 20 gram piece and one twice that weight you might reasonably have a strict preference.

If differences cease to be imperceptible, then a reason for preference may appear. Machina, whose version the chocolate-bar example is, suggests that successive declarations of apparent indifference might mask the strict preferences that would appear if to every piece of chocolate a label was attached stating its weight. I know of no evidence directly bearing on this proposition, though experiments that show positive preferences between identical items suggest how it might be violated. However, it is surely right not to leave the story as it is; for it seems to leave intransitivity of indifference accepted for reasons to do with bounded rationality. If this justification were to be accepted, then critics of SEU would seek similar warrants for violations of all axioms.

One of the earliest formal experiments on transitive preference[5] is reported in Rose (1957), who took it as given that intransitivities were a fact of life. However, he felt that this should not concern advocates of transitivity if intransitive choices could be judged not 'true'—that is, if they could be shown to be random. Seventy-four students of sociology were asked to rank the seriousness of 13 crimes in all their possible 78 pairs. Two pairs were repeated in each experimental session and a complete retest was conducted two months later. Rose wanted to interpret only those intransitivities that were repeated as being indicative of truly intransitive preferences. It was on that basis that he concluded that the hypothesis that all preferences were truly transitive could not be rejected.

An observer without Rose's declared desire that preferences should be transitive might interpret the results differently. For one thing, out of 74 subjects who took part, 71 made at least one intransitive judgement. Moreover, there were 24 subjects who repeatedly made an intransitive ranking. Not being able to reject a null hypothesis about nearly all preferences being transitive may be a correct way of describing these results, but it is partial too. The alternative reading is that nearly everyone has some intransitive preferences and that about a third of these appear to be systematic.

Rose is not alone in trying to down-play the relevance of random intransitivities, but he seems not to realize the double-edged nature of such a defence. If an error really is due to faulty response mechanisms that hide the real preferences, then these

---

[5] Earlier non-experimental work does exist; see e.g. May (1954).

errors will lead not only to bogus intransitive rankings but also to false transitive ones. If response error is some constant proportion of the total number of responses, then correcting Rose's data for such errors will lead to a larger number of changes from transitive responses to intransitive ones than the other way round. Taking the results at face value might lead to an underestimate of the extent to which preferences are intransitive.

Another experiment, conducted in the latter half of the 1950s and reported in Griswold and Luce (1962), tests a probabilistic version of utility theory devised by Luce (1958). Luce's theory was one of the first to explore the implications of stochastic axioms exemplified by this statement of probabilistic transitivity: if you prefer shredded wheat to muesli with a probability greater than one-half and you prefer muesli to corn flakes more than 50 per cent of the time, then, more likely than not, you also prefer shredded wheat to corn flakes. For better or worse, the literature has shown only sporadic empirical interest in stochastic axioms, and almost none at all in their normative status. Griswold and Luce's carefully designed experiment throws light also on the deterministic version of transitivity, and it is that part of the experiment which is of interest here.

Employing five subjects who were paid both a fixed sum and an amount contingent on their declared preferences, Griswold and Luce examined the nature of pairwise preferences between lotteries. In each session, subjects were offered lottery pairs that yielded either money prizes or different brands of cigarettes.[6] After a subject had chosen, the preferred lottery was played out on a randomizing device (a specially constructed pin-ball machine), and at the end of each session one of these plays was chosen randomly to determine the bonus part of the subject's payment. Three out of five subjects attended 50 sessions, while data for the other two were collected from 45 and 32 sessions.

While it is true that the thrust of the paper finds in favour of Luce's probabilistic theory, the data on transitivity tell a less straightforward story. When the gambles yielded monetary pay-

---

[6] *O tempora, o mores.*

Table 2.3 Tests and violations of transitive preference

Subject	No. of tests	No. of violations
1	2	2
2	0	0
3	3	3
4	2	0
5	13	1

offs, there was only one violation of the transitivity axiom although there were 56 cases in which a violation could have occurred. Choices between lotteries over cigarette packets indicate a different picture. Table 2.3 indicates, for each subject, the number of occasions on which transitivity could be tested and the number of times the assumption was violated. The assumption is testable on four subjects, and three of these violated the assumption at some stage. Furthermore, in the cases of two subjects all the possible violations that could be made were made. Although the number of subjects used is small, the intensive nature of the experiment means that subjects are more likely to understand the tasks. Nevertheless, two subjects were intransitive whenever they could be. It is interesting to speculate that the higher proportion of intransitive preferences may be due to the non-numerical content of the rewards in this experiment.

Possibly the best known experimental evidence against transitivity is in a meticulous investigation reported by Amos Tversky (1969). Tversky predicted that subjects would make pairwise comparisons between 'similar' lotteries on the basis of their highest pay-offs but that choice between substantially different lotteries would be on the basis of expected value. A number of intransitivities appeared and many were of the kind Tversky had predicted.

In addition to these three studies, the body of experimental work on transitivity includes experiments by Morrison (1962), Shepard (1964), Navarick and Fantino (1974), Montgomery (1977), Lindman and Lyons (1978), Haines and Ratchford (1987), Bradbury and Ross (1990), and Korhonen *et al.* (1990).

If one looks for it, evidence of intransitivity is not difficult to find, though there has been a greater tendency to acknowledge empirical failures of independence. One possible reason for this has to do with the academic climate in which experiments are interpreted. A comment by Rose helps to make the point:

> Traditionally economic theory has been based on the assumption that behaviour is 'rational' . . . whereas most psychological and sociological theory insists that behaviour is, at least, largely, irrational. . . . Some of the more advanced [non-social scientists] . . . have been discovering that much of what superficially appears to be 'irrational', in terms of what the outsider considers to be the interests of the subject or the information available to him, is really 'rational'. . . . (Rose 1957)

Rose then proceeds to note that many of the assumptions made by political economists have been shown to be unnecessary by latter-day econometricians but that rationality, to which transitivity was regarded as integral, was still essential. It was as if rationality and transitivity serve, for many economists, as a 'boundary object' which marks off their territory from that occupied by other social scientists. Had the earlier experimentalists been writing today, they may well have described their results in such a way that the intransitivities were emphasized, as, for instance, did Grether and Plott (1979).

COMPLETENESS

Whether preference is complete is particularly dependent on how we interpret the relation. The conventional view, even among advocates of SEU, is that completeness is an axiom with little normative appeal. Few experiments seek to test the assumption explicitly, though there are two that can be mentioned in this context.

The first study, conducted by Ken MacCrimmon (1965), is an investigation of all the axioms of SEU in a sample of experienced business executives. If completeness requires that decision-makers must have a preference or be indifferent between options, then a way of testing the assumption is to ascertain whether preferences change with alterations to the set of feasible options. If a person prefers *a* to *b* when only *a* and *b* are available but prefers *b* to *a* when *a*, *b*, and *c* are available, the 'choice instability' might be interpreted as a violation of completeness. Choice instabilities are not ubiquitous, though 50 per cent of all subjects exhibited them at some time.[7]

An alternative operationalization of completeness is suggested by the finding that on no occasion did any subject refuse to rank options. Had there been choices in which options could not be ordered with respect to preference, these too could have been taken as a violation of incompleteness. In contrast, it is not difficult to think of situations outside the laboratory where completeness is violated. You might 'have to' choose one out of two illegal or morally repugnant options and not want to say that you *preferred* either. More concretely, the number of abstaining voters in each UK parliamentary constituency who go to the effort of spoiling their ballot papers is often in the hundreds.

A second strand of evidence relating to the completeness of preferences comes from a series of experiments initiated by cognitive psychologists at the end of the 1960s and re-examined in the *American Economic Review* throughout the 1980s. This body of work deals with a phenomenon that has come to be known as 'preference reversal', which came to the attention of economists through a carefully designed experiment by Grether

---

[7] It might be argued that these are violations of the requirement that preference should be independent of other elements of the feasible set—the 'independence of irrelevant alternatives' assumption. This is not an explicit assumption of SEU and therefore its violation seems not to falsify the theory. However, most, if not all, proponents and users of SEU assume that the theory applies to problems where the decision-maker has two *or more* options. If it were not possible to think of preferences over two items as being extendable to decisions over three or more options, SEU would be applicable in very few of the instances where it has been used. The most obvious way in which the 'binary' preference relation of axiomatic accounts of SEU can be extended to choices over three or more options is to assume that *Pab* implies *Pacb*. (There are technical variations of this assumption, e.g. that *Pab* implies $\neg Pbca$, but for most normative purposes the variations are unimportant.) If such an extension is implied by the use for which SEU is advocated, then it is fair to test the assumption on which the extension is justified. Whether the test is described as a test of completeness or of another implicit assumption is not of major import. Both readings seem intuitively plausible.

and Plott (1979). There are distinct types of preference reversal, but they are all variants on a similar theme. First, subjects are asked whether they would rather play a lottery with a high probability of paying out (the 'p' bet) or one that yields a high pay-off with a smaller probability (the '$' bet). Then subjects are asked to say what minimum sum of money they would accept (known as the 'minimum reservation price') so that they would forgo the opportunity of playing the same lotteries. In experiment after experiment, a large number of subjects judge that the p bet is more desirable, though they put a higher minimum reservation price on the $ bet. If one allows that the direct preference question and the minimum reservation price are measures of desirability for the same thing, the interpretation of the reversal phenomenon as evidence of incompleteness follows.

The interpretation of preference reversal as a violation of completeness is but one interpretation. In their original paper, Grether and Plott took their results to be evidence against transitivity and therefore against the notion of maximization itself. Subsequent work has shown that reversals can be explained as a failure of experience (Pommerehne *et al.* 1982), independence (Karni and Safra 1987), or the reduction postulate (Segal 1988). One of the more recent contributions to this debate can be found in work by Tversky *et al.* (1990). Their paper reports two experiments conducted on some 198 students at the University of Oregon. I shall discuss only the first of these, though both yield broadly similar results.

The aim of the experiments was to distinguish between four possible causes of preference reversal. The first two relate to the way in which probabilities enter into the decision-maker's utility function. They seek to identify violations of independence and the reduction principle. (Like independence, the reduction principle is related to the expectation operator and serves to produce a single lottery from lotteries where uncertainty is resolved in two or more stages. For example, a 50 per cent chance of going for a jackpot that pays out one time in ten is, in reduced form, taken to be equivalent to a 5 per cent chance of winning the jackpot.) In addition, they were able to distinguish between reversals due either to intransitive preferences or to a further assumption which they call 'procedure invariance'.

To test the possibility that failures in independence and/or reduction were responsible for the reversals, two methods were used to elicit preferences. One of these relied on the assumption of independence and the other on reduction so that reversals due to each would show up as differences in group comparisons. The fact that the proportion of reversals was broadly the same, regardless of which method was used, eliminates violations of independence and reduction as possible explanations.

To distinguish between the other two possibilities, intransitivity and procedure invariance, subjects were asked to make choices from the sets {p bet, $ bet}, {p bet, cash}, and {$ bet, cash}, where 'cash' stands for a certain financial payment. Subjects were also asked to give valuations of the p and $ bets: responses used for further analysis were those in which the ordering of bets by value yielded p bet > $ bet. If the choices lead to the rankings $ bet > cash, and cash > p bet, it would seem that the preferences are indeed intransitive. Three other situations are possible: $ bet > cash, and cash < p bet; $ bet < cash, and cash > p bet; and $ bet < cash, and cash < p bet. None of these patterns combines with the higher valuation for the p bet to yield an intransitive preference, so that these three preference patterns, if they emerge, can be used to eliminate the intransitivity line of explanation. In fact, in both experiments some 90 per cent of responses fell into one of these three categories, suggesting not that transitivity is violated but that most preference reversals constitute a violation of scale invariance. Tversky et al. (1988) provide a theory[8] that explains such bias in terms of the similarity of the response mode with the description of the gambles. $ bets seem more desirable in valuation tasks than in preference-ranking questions because the request for a numerical dollar response causes subjects to place more weight on the elements of the lottery in similar units

[8] Goldstein and Einhorn (1987) also provide a theory that links responses to the modes in which they must be given. The underlying theme, that responses are contingent on the output required from the evaluation task, is one that can be traced back at least to Tversky (1969).

(i.e. the pay-offs). These findings may have no normative impact, but that does not detract from their significance as empirical violations of SEU and many more general theories of decision.

Following Frank Knight (1921), economists often distinguished between risk, where relative frequencies exist, and uncertainty, where they do not. The philosophical distinction between aleatoric uncertainty, arising from randomness, and epistemic uncertainty, which reflects a lack of knowledge, is not unrelated. If uncertainty has any significance, this should make itself apparent in methods for analysing decisions taken in the absence of probabilistic information. Two possible views have attracted support among economists. One tradition, known as decision-making under ignorance, focuses on the rules that deciders can apply to the consequences of different actions. In the absence of knowledge about probabilities of consequences, rules like 'maximin' (pick the act that has the best, worst pay-off) emerge. Theories of this kind are enjoying something of a revival at present, though little use has been made of such theories in the general economics literature. The approach that gained greater support was that proposed by the subjective probabilitists. The case they make is that, even in situations of uncertainty where no objective probabilities exist, subjective ones can be inferred from a person's behaviour.

This view was first shown to be incorrect in an experimental paper due to Daniel Ellsberg (1961). Like Allais's paradox, Ellsberg's results are presented as a violation of an assumption in Savage's axiomatization, although the probabilities he assigns to states are not the usual point estimates but are defined only up to an interval. The experiment discussed here is one of two that Ellsberg describes; details of the experiment are summarized in Table 2.4.

The subject must choose from an urn in which there are 30 red balls and 60 black and yellow balls in unknown proportion. The choice is between an act A, which yields $1,000 if a red ball

Table 2.4   Ellsberg's uncertainty problem

Acts	Number of balls		
	Red	Black	Yellow
	30	60	
A	$1,000	$0	$0
B	$0	$1,000	$0
A'	$1,000	$0	$1,000
B'	$0	$1,000	$1,000

is selected, and act *B*, which yields the same monetary sum if a black ball is picked. The second choice offered is between *A'* and *B'*, which differ from *A* and *B* by offering $1,000 if a yellow ball is picked. Independence requires that strict preference for *A* (*B*) should accompany strict preference for *A'* (*B'*), though the most frequently observed choice pattern was *A* and *B'*. Ellsberg suggested this was evidence that a significant number of subjects were averse to uncertainty and preferred prospects associated with point-defined probabilities to those for which only probability ranges could be specified.

Roberts (1963) offers a methodological critique of the experiment which casts doubt on the inference that Ellsberg draws. Subjects might have picked the 'safe' options *A* and *B'* in fear of the inferences that an experimenter might make about them if they did otherwise. Alternatively, they might have believed, albeit wrongly, that the pay-off was related to the probability of winning. If such subjects thought that the prize resulting from *B* would be something between nothing and 60/90 of $1,000, say, then risk aversion could explain the observed choices. The basic probabilistic analysis of the situation might be misunderstood, or the subject might just be indifferent and express the first preference that comes into his or her head. Ellsberg (1963) provides a robust, though theoretical, rejoinder to the problems Roberts raises. An attempt to control for these concerns appears in Anand (1986, 1990*c*). In this version of the experiment, subjects were asked not just to express a preference but to

Table 2.5 Reasons for choice under uncertainty

Reasons	No. of responses
Uncertainty avoidance	12
Uncertainty preference	4
Indifferent	7
Instinct	2
Don't know	4
Misinterpreted question	4
Total	33

indicate their understanding of the statistical nature of the problem and their reasons for choosing: 30 students (mainly undergraduates) were asked to indicate how they would choose if the options were (*a*) win £100 on a black ball from an urn containing 5 out of 10 black balls and (*b*) win £100 by picking a black ball from one of two unmarked urns. Both urns contained ten balls: in one seven were black, in the other three were black.

In part, the results give some support for Roberts' concerns. Only 21 subjects indicated that the probability of picking a black ball and winning the prize from the unmarked urn is 0.5. Responses to the question about reasons for choice were categorized (by the experimenter[9]), and are summarized in Table 2.5. Some subjects gave responses that fell into more than one category.

Again, there is some support for Roberts' objections. Subjects did not know why they had made a choice, or indicated that, despite having expressed a strict preference which was unforced, they really were indifferent between options. Nevertheless, over 50 per cent of the sample (16 subjects) understood the problem, made a positive choice, and gave an uncertainty-related reason for choosing that was choice-consistent. It is this triangulation finding that seems ultimately to support Ellsberg's uncertainty aversion; indeed, there are a number of experiments dealing with behaviour under uncertainty (sometimes called 'ambiguity' or

---

[9] For a justification of this procedure, see Anand (1990*c*: 152–3).

'vagueness') that confirm the extensiveness of what might be called uncertainty aversion—see for example Becker and Brownson (1964), Yates and Zukowski (1976), Goldsmith and Sahlin (undated), Einhorn and Hogarth (1986), and an excellent field study by Steil (1993). The value of Ellsberg's work lies, of course, in the technical point he makes about the incompatibility of uncertainty aversion and the existence of a subjective probability.

VIOLATIONS OF COMMON INTERPRETATIONS OF SEU

As Savage points out, if SEU is to be successful, the consequences in the act–state matrix must be complete descriptions of everything that matters to the decision-maker. Perhaps this comprehensiveness requirement seems overly permissive, but by observing the way SEU is used it is possible to infer interpretational constraints that give the theory testable content. None of the evidence considered in this section deals with axiom violations *per se*, although all are inconsistent with usual readings of SEU.

Reference Point Effects

The use of reference points in axiomatic decision theory can be traced back to work by Shackle (1961). Although the focal points of Schelling (1960), which they resemble, have for a long time interested game theorists, reference points were not widely discussed in decision theory until the 1980s. It is known that attitudes to risk typically change from aversion to preference depending on whether the domain consists of gains or losses. Such preferences can be represented by a utility function that has both a convex and a concave section. The point where these sections meet is a reference point. While this is consistent with the idea of a utility function (assuming the join is continuous), it plays no substantive role in SEU.

Evidence that indicates the significance of such points can be found in at least two literatures. In the first place, economists implementing social cost–benefit analysis have noted that valua-

Table 2.6   Valuation disparities

	Yes	No	Total
WTP > $2	19	19	38
MRP > $2	29	9	38

tions of commodities vary considerably depending on how the valuation question is put. When respondents are asked to say how much they would be willing to pay (WTP) for a commodity, they give a valuation that is much less than what they say their minimum reservation price (MRP) would be for the same item.[10]

It is often argued that these responses reflect the opening positions that people would take in a bargaining situation, though an investigation that seems to eliminate the possibility appears in Knetsch and Sinden (1984). In their experiments, subjects were asked only whether their WTP or MRP values for a lottery with certain prizes were more, or less, than $2. Given their design, the question seems to leave little ineliminable opportunity for giving an untruthful response. (The question is 'incentive-compatible'.) Their results, summarized in Table 2.6, show that the WTP and MRP valuations are different (at the 5 per cent level of statistical significance).

The second source of support for reference-point phenomena can be found in the experimental social psychology literature. In several experiments, Kahneman and Tversky (1979) have shown that even the way in which final outcomes are described can determine the modal attitude to risk. (On this, see also Arrow 1982.) The following choice situations help illustrate the issue:

Situation 1: In addition to whatever you own, you have been given £1,000. You are now asked to choose between:

| Option *A*: | $p = 0.5$ of winning £1,000 | (16%) |
| Option *B*: | $p = 1$ of winning £500 | (84%) |

Situation 2: In addition to what you own, you have been given £2,000. You are now asked to choose between:

---

[10] Results in Anand (1986) show that a fourfold difference can be average.

Option A': $p = 0.5$ of losing £1,000   (69%)
Option B': $p = 1$ of losing £500   (31%)

When viewed in terms of end-wealth states, a natural frame to use, the choices in situations 1 and 2 are identical. However, as the figures in parentheses show, the proportion of subjects picking the lottery substantially increased when the incremental parts of the gambles were negative. We noted before that attitudes to risk depend on whether the pay-offs are positive or negative. The fact that the same lottery can be described in either way suggests that, at least for predictive purposes, one needs information on the way in which the problem is conceived by the decision-maker. SEU is not normally associated with this kind of information.

Context Effects

While it is a truism to say that behaviour is affected by the context in which it takes place, domain specificity is not an issue easily dealt with by SEU. The practice of assuming the existence of a utility of money function for use in decision-making suggests that, tacitly at least, attitudes to money risks can be abstracted from the social domains in which those risks are borne. Schoemaker and Kunreuther (1979) and Hershey and Schoemaker (1980) report such experiments. One of these provides subjects with lotteries that are identical at the level of money pay-offs but are so formulated that they differ in social context:

Formulation 1: Choice framed as a gamble
   Gamble 1:   A sure loss of $10
   Gamble 2:   1% chance of $1,000 loss

Formulation 2: Choice framed as an insurance decision
   Option 1:   Pay an insurance premium of $10
   Option 2:   Remain exposed to a 1% loss of $1,000

Of 42 subjects, 51 per cent chose the sure loss in the gamble formulation, while the figure rose to 81 per cent when the loss was described as an insurance premium. How these differences could naturally be modelled by SEU is not clear.

Evidential Reasoning

In the course of an elegant reformulation of Ramsey's original version of SEU, Jeffrey (1965) proposed that the primitives of decision theory should be propositions. The use of propositions in place of acts, states, and consequences reduces considerably the kinds of primitives required, but leads to a question about what the utility of a proposition could be. Jeffrey suggested that this might be the news value to an agent of knowing the proposition is true, a device he attributed to Savage.

In the late 1960s, Newcomb proposed a problem which seemed to show that news value maximization leads to normatively incorrect choice. To the scientist the original problem is fanciful, but subsequent work has shown that realistic applications do exist.[11] The following version is based on Newcomb's original cover story:

> There are two boxes and you can choose either from one, or from both. One is sufficiently transparent that you can see that it contains £1,000 before you choose. The other is opaque and the decision to pick from it must be made before its contents are known. Furthermore, there exists a super-being who will guess how you are going to act and may fill the opaque box depending on his prediction about your choice. If the super-being thinks you will only choose from the opaque box, then he places £1m in it. If, on the other hand, he believes you are going to choose from both boxes, he will leave the opaque box empty. (see Nozick, 1969)

You can choose to pick either from the opaque box or from both the opaque and the transparent boxes. The questions here are which does SEU suggest, and how do subjects actually behave in practice? The 'causal' analysis, to which most users of SEU would subscribe, runs thus. As the super-being has moved first, the dominant choice is to pick from both boxes. That way you get either £1,000 or £1.001m depending on the prediction the super-being has made about your choice. To do otherwise, that is to pick from the opaque box only, is to take a chance that will yield £1m if the super-being guessed correctly and

---

[11] See e.g. Cubitt (1986), or Broome (1989).

nothing if he did not. The alternative, 'evidential', view holds that, because picking one box indicates a higher probability that the opaque box has been chosen, this can still have a higher expected value despite its lower pay-offs. This is dismissed as an irrelevance by the causal school. Whether the opaque box is full or not, the super-being has already moved so you might as well pick from both boxes.

Explicitly, this issue has been discussed mainly in the philosophical literature. However, one of the first things that students learn in economics is that fixed costs should have no bearing on current actions—'bygones are bygones'; implicitly, most economists are causal decision theorists. What people do in practice is, of course, another matter, and experiments reported in Anand (1990c) suggest that evidential choosing is widespread.[12] According to the causal analysis, 'two-boxing' is dominant so changes in certain parameters, for instance the predictive ability of the super-being, should have no effect. In contrast, evidential choosing predicts that the attractiveness of one-boxing can be made to vary and indicates how it will do so. For example, the bigger the large pay-off might be, the higher the evidential expected utility of one-boxing. In a number of tests it was found that subjects did pick one-box choices, and the proportion of subjects doing so varies both with the super-being's predictive ability and with the value of the possible pay-off in the opaque box. Realistic versions were also used to test whether the result was an artefact of the abstract nature of the problem. The change appears to make little difference, as up to half the subjects made evidential choices, even when the cover story was realistic.

CONCLUSION

Most sciences concern themselves with the way the world is. SEU is one of many theories that compete for our attention as an account of the way decision-makers behave, and it has

---

[12] Methodological details are discussed in the paper.

played a vital role in the development of decision theory and economics. This chapter reviewed a fraction of the empirical evidence relating to choice under uncertainty. Some of it would not exist were it not for the axiomatic structure of SEU, but very little seems to validate SEU. Rather, the evidence that does exist seems to go against both explicit and implicit assumptions. This should not surprise die-hard supporters of SEU, as even simple choices which should be based on safe first logic show no signs of being so.

Economics, especially mathematical economics, shares with philosophy a number of normative goals. Whether these goals are scientific or not is disputable, but the point is that normative goals are, as a matter of fact, widespread in social science. Perhaps the fact is not everywhere recognized, but it is a key to understanding the case for SEU. The next chapter, therefore, seeks to clarify what the goals of decision theory are.

3

Theory Aims and Evaluation

INTRODUCTION

We have seen how SEU satisfies a need for a subjective theory of decision. All the substantive axioms have been empirically tested, however, and it is difficult to see how one could argue for SEU on empirical grounds. On the other hand, falsification need not lead to rejection if scientists have other uses for their theories. For any evaluation of SEU to be fair, therefore, we should understand the objectives that proponents of SEU have set themselves. The role that SEU played in providing foundations for much of economic analysis means, almost inevitably, that we are obliged to touch on issues of economic methodology. Indeed, there are a number of reasons why neo-classicists could, legitimately, insulate SEU against evidence of empirical falsification.

In this chapter, three distinct methodological phases in the history of utility theory are identified. Each phase differs with respect to the goals that it pursues. Sceptical readers of Kuhn might say that each is just another attempt to immunize SEU, but there is no reason why the objectives of science should remain immutable. The current agenda of decision theory emerges as a clarification of long-standing concerns.

THE PRE-HUMEAN FOG

Philosophers often attribute the distinction between what is and what should be to David Hume. Work associated with the Enlightenment origins of decision theory exhibits an awareness of this distinction, though without the sharpness Hume gave to it. In the treatise in which he solves the St Petersburg paradox,

Bernoulli (1739) wrote that 'any fairly reasonable man would sell his chance to play the St Petersburg paradox, with great pleasure, for twenty ducats'. Gabriel Cramer seems to have been motivated by similar considerations when he writes:

I believe that it results from the fact that, in their theory, mathematicians evaluate in proportion to its quantity while, in practice, people with common sense evaluate money in proportion to the utility they can obtain from it. (Cramer 1728)

Both men were concerned, then, by what they took to be empirical observations about reasonable people. This is not to say that they confused normative and positive issues, but it does indicate a willingness to let them be confounded. Even Ramsey echoes this position. He associates subjective probability with consistency, valued for the protection it offers the agent from smart bookmakers offering lotteries ('Dutch books') that lead to certain loss. Yet he adds to gloss that each of the definitions used in his account of SEU is implicitly 'accompanied by an axiom of consistency, and in so far as this is false, the notion of the corresponding degree of belief becomes invalid'. To call this stage a pre-Humean fog is just to say that the early history of decision does not let the is–ought distinction play a substantial role in shaping theory. Nevertheless, both issues are important.

THE ALL-IN-ONE VIEW

In their account of SEU, von Neumann and Morgenstern briefly discuss criteria that could be used to construct, or evaluate, axiomatic theories. They write:

The axioms should not be too numerous; their system is to be as simple and transparent as possible, and each axiom should have an immediate intuitive meaning by which its appropriateness may be judged directly. (von Neumann and Morgenstern 1944)

The thesis seems to be that theory evaluation is not some holistic, Gestalt process. Rather, the task of determining the appeal of a theory can be decomposed into a series of individual axiom evaluations.

The comments von Neumann and Morgenstern make about individual axioms add further insight into how the business of axiom evaluation is conducted. Of transitivity, for example, they suggest that it is 'plausible' and 'generally accepted'. Other axioms are defended on more pragmatic grounds. Regarding completeness, they allow that 'one may doubt whether a person can always decide', but note that the assumption is required by the analysis of indifference curves (the main competitor of utility theory at the time). Again, they remark that, while the intrinsic pleasures of gambling may give grounds for violation of independence, this cannot be 'formulated free of contradiction' at the axiomatic level. Independence, according to von Neumann and Morgenstern, is justified not by its empirical validity, but rather by the needs of the axiomatic method.

This loose approach to axiom justification is reasonable so long as we recall the context in which the axioms were proposed. From one perspective, this is the technical experiment that paves the way for subsequent uses of axiomatic measure theory in a number of areas of economics. However, given the maturity of the theory, and in comparison with the technical aspects of utility theory, it is easy to be dissatisfied with the arguments for the axioms themselves. The warrants range from the mathematical needs of axiomatic system formulation through to social convention and personal opinion. The fact that very different kinds of justification are used to defend each axiom makes it difficult to establish a single *raison d'être* for the theory.

In a major review of the history of utility theory, Stigler (1950) identifies three criteria that characterize improvement, not of assumptions, but of theories. This 'theory of economic theories' applies to economics in general, but its application to utility theory is what interests us here. First, there is a requirement that new theories be more general than those they seek to supplant. A—if not *the*—major benefit of generality is that any 'results' (analytical consequences) of more general theories are applicable in a wider set of circumstances. With the exception of Marshall's assumption of constant marginal utility, Stigler claims that there was no 'important instance' of a more specific economic theory replacing a more general one. The importance he attaches to it, contrasted with the notion of science as *narrowing* possibilities by eliminating false hypotheses, serves to emphasize just how much closer the dynamics of paradigm development in economics are to those of pure mathematics than to those of Popperian natural science.

A second criterion identified is manageability—the requirement that deduction from, and manipulation of, theory be easy. Currently, the requirement that theories be tractable is most often used to defend certain applications of mathematics to economic theory, although Stigler suggests that there was a time when the opposite was true:

Economists long delayed in accepting the generalised utility function because of the complications in its mathematical analysis, although no one (except Marshall) questioned its realism. (Stigler 1950)

The third criterion concerns the extent to which utility theory appears consistent with the facts. That Stigler places empirical validity third reflects the role that facts seem to play in the development of economic theory. Non-economists may think this strange (a physicist once remarked to me that economic theory should be more physics-like in the sense that it should be more empirical), but there are arguments for both approaches. In any case, and based on developments in utility theory in the 1980s, at least some areas of economic theory have shown that they are, over the long haul, sensitive to new empirical findings. The circular relation between the new theories of decision-making and experimental findings already shares the methodology of modern physics.

A position shared by Stigler and von Neumann and Morgenstern is that the theory which predominates does best according to a number of criteria. Stigler's article is surprising in that it says nothing about the importance of rationality. It may be that this is a modern concern, though I am more inclined to see the omission as evidence of a disparity between the business of economics and our attempted reconstructions of what economists do.

In the 1980s, a number of philosophers of science proposed

pragmatic theories of explanation (e.g. Achinstein 1983). According to these views, whether something is explanatory depends critically on what kind of question is being asked and who the audience is[1]. At about the same time, the wealth of experimental work on SEU caused decision theorists to wonder if it was not asking too much to expect one theory to serve both positive and normative needs. From very different academic traditions, both strands of thought argue for a more contingent approach to theory evaluation.

THE PROBLEM-CONTINGENT VIEW

There is a growing tendency to distinguish between the mathematical aspects of a theory and its interpretations. A hyperformalist position is taken by Debreu (1959), who equates theory with its mathematical parts. Others take a different view. Quine (1981), for instance, suggests that even pure mathematics is never '*dis*interpreted'. Whatever one thinks a theory is, we can evaluate an abstract calculus and its models separately, and we now consider the three bases on which such an evaluation might be conducted.

Positive Theory

Best understood are the positive aims of prediction and explanation that decision theory shares with natural sciences. An interesting explanatory use is provided by Friedman and Savage (1948), who show that an S-shaped utility function (which is concave but becomes convex) can explain why poor people both take out insurance and buy lottery tickets. Most advocates of SEU, however, place great importance on prediction, although, as we saw in Chapter 2, it is difficult to justify SEU on the basis of the predictions its axioms lead one to make.

---

[1] See also Federkke (1993) for an application to economics.

Conditional and Partially Interpreted Theory

Decision theory has for some time shared with economics aims that are only just beginning to be recognized. Friedman's claim that utility maximization should be read metaphorically (the 'as if' proposal) was part of an instrumentalist methodology he espoused. Theories should be evaluated not by their assumptions, he argued, but rather by the outcomes they lead an observer to predict. The long history of axiom testing in utility theory indicates that these views were not ubiquitous, even in economics. Some researchers now claim virtually the opposite: namely that their theories are (just) descriptive.

Given that a common rationale for theory is the escape from description that it offers, the idea of a descriptive theory is not entirely straightforward. Nevertheless, the purported aim of such theory—to capture observed behaviour within an axiom set—is clear enough. As we noted, the pressure to provide more general theories—which the descriptive project implies—would seem to be moving in the opposite direction to that demanded by the task of making falsifiable predictions.

An alternative, Debrovian, approach is to say that the mathematical components are conditional, linguistic theories. If certain conditions hold, then a particular description (of certain choice behaviour) follows. These theories are moodless. They can be used to make descriptive statements, issue commands, give advice, and make judgements, but the evaluation of such acts should not be equated with the evaluation of the theory. Like the descriptive view, being able to describe behaviour in a particular way is important, though under the conditional view facts about choice behaviour do not, directly, matter. In the first case, generalizations of SEU are improvements because they capture more data. Under the formal view, SEU is just one of a growing set of representation and uniqueness theorems in which all are equals.

A family tree which provides a taxonomy of developing areas in decision theory is depicted in Fig. 3. It was noted in Chapter 1 how the origins of SEU can be traced back to the use of the expected-value rule. SEU itself is now a springboard for the new decision mathematics that emerged in the 1980s, although some of these results predate the decade.

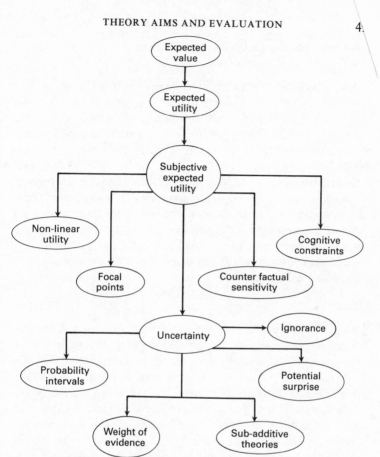

Fig. 3   Mathematical theories of decision-making

Counterfactual Sensitivity

The first theory to drop the transitivity assumption can be found in the additive difference model due to Morrison (1962), further discussed by Tversky (1969), and axiomatized in Fishburn (1981), and Croon (1984). In this model, the consequences of an act are compared with those of the alternative option. Option differences are evaluated for each attribute and then aggregated to provide an overall judgement. Formally, the agent prefers a prospect x to prospect y if and only if

$$\Sigma f[u(x) - u(y)] \geq 0,$$

where $u(x) - u(y)$ is the utility difference between x and y on a particular attribute, f is some function of those differences, and summation takes place over attributes. Morrison's ideas here are cast in terms of multi-attribute theory without explicit reference to uncertainty. More recently a number of writers, including Bell (1982), Fishburn (1982), Loomes and Sugden (1982), and Anand (1982), have analysed models with related mathematical structures interpreted as theories of choice based on a single attribute and many possible states. As these models seek to capture concern for what might have happened if the world had turned out differently (disappointment) or if the decision-maker had chosen differently (regret), they fall into the class of counter-factually sensitive decision theories.

Focal Points

SEU makes no explicit allowance for the possibility that certain consequences might come to mind more readily than others, though other theories do this. The economist George Shackle (1961) was the first to exploit the idea of focal gains and losses in an axiomatic theory of potential surprise (an alternative to probability), although it is the reference points of prospect theory due to Kahneman and Tversky (1979) that brought salience to the fore. The experimental literature on the subject (see for instance the review in Nisbett and Ross 1980) is substantial, and theoretical developments in the context of game theory can be found in Bacharach (1991).

Cognitive Constraints

This category of work is included for completeness rather than for the impact such models have had on decision theory. Mathematical theories based on bounded rationality in Simon (1957) are now used by economists, though experimental psychologists have a long tradition of modelling human cognition (see e.g. Restle and Merryman 1969). In decision theory, these models are only beginning to be employed. One that does this is expression theory (Goldstein and Einhorn 1987), which claims that the observed measures of preference are affected by the cognitive process used to evaluate the gamble, and that this, in turn, is a function of the mode in which subjects are requested to respond. In expression theory, Goldstein and Einhorn formulate the desirability of a gamble, $u(G)$, as given by the equation, $u(G) = u(win) - p[u(win) - u(lose)]$, where *win* and *lose* are the two possible outcomes, u is a function measuring desirability, and p is a parameter reflecting the extent to which subjects anchor evaluations on the lower outcome. The minimum reservation price, $mr(G)$, is given by the expression $mr(G) = win - f(p)(win - lose)$, where $f(p)$ indicates an anchoring process different from that used in judgement tasks. Using these different rules, it is possible to model a variety of preference reversal phenomena, and the theory of contingent weighting due to Tversky *et al.* (1988) achieves similar results.

Non-Linear Utility

The literature that generalizes expected utility to functions that are not linear in probability is particularly extensive. Knowledge of, and interest in, this work owes much to the efforts of Mark Machina (1982, 1987) and can be traced through to work by Karmarker (1978) and even earlier research on models of preferences for particular probabilities due to Phillips and Edwards (1966)[1]. Karmarker proposes a non-linear generalization of the expected utility rule in which the subject maximizes the value of $\Sigma p'u/\Sigma p'$, where p' reflects a subjective weighting of a probability given by the function

$$p' = [p/(1 - p)^{\alpha}]/[1 + (p/(1 - p)^{\alpha}].$$

The commonly observed situation in which small probabilities are over-weighted and large probabilities are under-weighted corresponds to the parameter α taking a value of less than one. Other non-linear models can be found in Handa (1977), Kahneman and Tversky (1979), Quiggan (1982), Chew *et al.* (1987), and Luce (1988).

[1] Maurice Allais's contribution, mentioned in Ch. 2, should not be forgotten.

Uncertainty

With the exception of Shackle's work, the theories mentioned so far resemble SEU in that they analyse uncertainty within the framework provided by probability. Other ways of conceptualizing credence do exist, and mathematical attention has been focused on these also. Theories with axiomatic foundations, motivated by attempts to describe uncertainty aversion found in Ellsberg decision problems, appear in Gilboa (1986), Schmeidler (1989) and Nakamura (1990). They represent a sub-additive approach to expected utility—so called because the agent maximizes Σpu but without the constraint that $\Sigma p = 1$.

A theory by Sarin and Winkler (1990) examines choice over probability intervals while others dispense with probabilities altogether. These theories used to be dismissed as imputing to human beings a level of ignorance that would virtually never be encountered in the real world, but such theories are once again being examined (see Arrow and Hurwicz 1972; Cohen and Jaffray 1980; Barbera and Pattanaik 1986; Kelsey and Quiggan 1989). Still other theories develop new concepts of credence. Potential surprise, discussed by Shackle (1961) and Levi (1966), has already been mentioned, and a similar possibility theory is axiomatized by Dubois (undated). Keynes (1921) also proposed a non-probabilistic measure of the quantity of information which he called the 'weight of evidence'.

Most classes of models discussed above are known to contain a number of theories that generalize SEU. For a technical guide to some of these, Fishburn (1988*a*) is difficult to better. In a purely technical light, SEU is one of many theories of decision and it is certainly not the most general theory. However, most researchers would allow that decision theories have to have some model. We saw that there are problems with the empirical interpretation, and that seems to leave only one non-pragmatic source of justification.

Normative Theory

It would be strange if decision theory were incapable of helping decision-makers. In fact, it can help them in two ways. In the first place, it could be used in a literal way as a prescriptive aid to deliberation. Decision-analysts have devoted considerable effort to the task of making decision-theory usable and there is a considerable literature just on this—see e.g. von Winterfeldt and Edwards (1986). Second, some of the benefits reported have less to do with the application of rigorous technique and relate more to the new exchange of information and social consensus that seem to follow from careful application of decision-analytic techniques. For those who advocate the prescriptive use of SEU in such contexts, there is little opposition and much support.

This first, practical, use does not however underwrite what economists mean when they say that SEU is normatively desirable. The alternative interpretation of the normative project, sometimes confused with that outlined above, seeks to provide an answer to the following question:

(Q1): How should an ideal agent choose?

A necessary part of any answer to this question is a decision about how best to cast the ideal agent. For some political economists, an ideal agent was a person sufficiently educated to know what was in his or her own best interests, with sufficient will-power to act upon those interests. For many modern purposes, it suffices to say that the ideal agent has perfect, instantaneous, and costless powers of calculation. It might be said that modern answers to the question above must satisfy the following constraint:

(C1): Solutions to (Q1) should be based on the maximization of consequences, taking into account (only) a person's beliefs and preferences.

This formulation leaves open, as is conventional in economics, the issue of what beliefs and preferences a person should have. Modern theories of rational action are at one and the same time consequentialist and subjective. The idea of a 'good' life would not seem to be part of such theorizing.

Solutions to (Q1) have been judged acceptable, typically, if they lead to choices that can be described as the maximization of a function. In fact, there are hardly any modern theories of idealized behaviour that do not provide theorems establishing the kinds of representation that certain behaviour permits. The requirement to provide such theorems essentially ties developments in utility theory to the branch of mathematics known as measure theory. This has proved a fruitful association, but it is not self-evidently the case that an ideal agent should behave in a way that is interesting from a measure-theoretic viewpoint. This is an extraordinary, bold contribution to a philosophical debate about the nature of rationality that dates back over 2,000 years. As we have seen, the normative warrant would appear to be the only substantial justification remaining for SEU as the main mathematical theory of decision. We have, it seems, a mandate to examine the claim, though the arguments in the following three chapters suggest that the defences of this Platonic bastion are not so sure as once we thought they were.

4

In Praise of Intransitive Preference

INTRODUCTION

Sometimes seen as the touchstone of formal theories of rational choice, transitivity is a puzzle. Of all the constraints that bind the preference relation, it was thought to be the most attractive, despite the paucity of firm arguments in its favour. Making the assumption used to be essential for the conduct of interesting mathematical theories of economic behaviour. Today this is no longer true: if anything, the converse is. Insisting on transitivity would rule out many of the new theories of utility maximization that decision theorists have shown to exist. From a technical viewpoint, it is anything but the progressive assumption it once was. Whether empirical violations are as extensive as those of independence depends largely on how preference reversals are interpreted, but in any case violations do exist. Is such behaviour irrational? Are models that drop transitivity necessarily models of irrationality? The conclusion to be drawn from the normative case discussed in this chapter is that we should answer these questions in the negative.

INTRANSITIVITY AS INCONSISTENCY: THE LOGICAL CASE

There is a sense in which intransitive preferences might appear to be a rather profound form of error. At first blush transitivity seems to be just a matter of logic, and it is tempting to think that its absence would constitute a case of logical inconsistency. Any observed violations would be judged as irrational in much the same way that we take behaviour based on incorrect syllogistic inference to be. The very title of Georg von Wright's *The*

Logic of Preference (1963) makes this association explicit, as does Broome (1991), through one of the earliest explorations of this position appears in Tullock:

The proof of intransitivity is a simple example of *reductio ad absurdum*. If the individual is alleged to prefer *A* to *B*, *B* to *C*, and *C* to *A*, we can enquire which he would prefer from the collection of *A*, *B*, and *C*. *Ex-hypothesi* he must prefer one; say he prefers *A* to *B* or *C*. This however contradicts the statement that he prefers *C* to *A*, and hence the alleged intransitivity must be false. (Tullock 1964: 403)

If the agent is assumed to prefer *B* or *C*, then a similar sort of contradiction can be produced. But what really has been proven? Let us revert to the use of lower case for choice objects and upper case for preference relations. We assume that preferences are strict, though nothing of substance depends on this. It seems we have a proof which says that any choice from the feasible set $\{a, b, c\}$ contradicts, i.e. is the negation of, a statement in the set of possible binary preference rankings (use *B* for binary preference), $\{Bab, Bbc, Bac\}$. Strictly speaking, any choice from the threesome must indicate something only about a *ternary* relation (which will be denoted by the predicate T^1) and would therefore [2] belong to $\{Tabc, Tacb, Tbca, Tbac, Tcab, Tcba\}$, none of which is an element of $\{\neg Bab, \neg Bbc, \neg Bac\}$, though one must be, if the contradiction is to be obtained.

The problem is that the attempted proof confounds two-place and three-place predicates. In general, these relations are different: '*m* divided by *n*' and 'Lucifer is the friend of Eve' are propositions involving binary relations, while '*r* is the integral part of *m* divided by *n*' and 'Lucifer is the mutual friend of Eve and Adam' both exemplify ternary relations. The binary and ternary relations are not unrelated but they are distinct, and this is true for all relations, binary and ternary preferences included.

If it is natural to use ' > ' as a symbol and analogy for strict

---

[1] To be read as in the binary case; e.g., *Tabc* should be read '*a* is preferred to *b* is preferred to *c*'.

[2] Embedded in the 'therefore' is a completeness assumption; nothing hangs on this, but assuming it makes what drives the argument more transparent.

preference, then perhaps it is easy to think that what is to be understood, in this case preference, has the same properties as the analogue. The value of predicate notation is that it reminds us that the chaining of relations that transitivity expresses is a matter of choice, not a given.[3] To pretend otherwise is to commit the fallacy of question-begging. None the less, it may be that a link between binary and ternary preferences can be constructed, and that this link serves the purpose the existence of which is being presupposed. Let us, therefore, examine the plausibility of such a link.

A principle that would complete the proof is one known as 'contraction consistency'. Founded on the idea that preference should not be affected if other items are removed from the choice set, the assumption can be stated formally (in logical notation $\Rightarrow$ is read 'implies that', and $\wedge$ is read 'and') thus:

$$(a)(b)(c)Tabc \Rightarrow Bab \wedge Bbc \wedge Bac.$$

Contraction consistency is particularly useful to an observer as it allows inferences to be made about n-ary preferences from knowledge about any higher-order level preferences. Transitivity does not imply contraction consistency, though it is a direct consequence.

If we force preferences to be bound by the contraction constraint, intransitivities are logically impossible. In other words, the normative appeal of transitivity seems to depend on our acceptance of the contraction constraint, though this is not an obvious result. The normative question can be shifted but not eradicated: do we believe that all agents should arrange their affairs so as to make sure their preferences satisfy the contraction constraint if they are not to be judged irrational? The answer turns, I suggest, on the nature of the decision problem that

---

[3] Even though the analysis here requires more precision than is customary in the interpretation of mathematical axioms, there is a certain looseness. To be precise, one would say that transitivity has nothing to do with moving between relations of different orders, but merely with moving between preference relations of the same order. However, the pedant would then have to agree that most axiomatic, mathematical theories of decision-making apply to choice between two options because they use only a single, binary preference relation.

underpins our preferences. In some cases goods can be regarded as having objective quantities of something that is desired. In such situations, it is hard to imagine circumstances in which the properties of an object might change depending on whether other objects were feasible choices. A packet of coffee beans, for instance, remains the same whether the store sells cheese or not. However, there are other problems for which context does matter, and in some of these the feasibility of certain options is a determinant of that context. To see this, and its force in undermining contraction, consider the following social choice problem.[4]

Suppose that you are asked by an opinion pollster to say whether you think patients should normally be allowed to see their medical records, and that your preference is as follows (> denotes strict preference):

(1)   no access to records > access to records.

You might feel that, if patients could see anything and everything that health workers wrote about them, members of the medical profession might spend much of their time, and taxpayers' money, responding to requests for access to medical records instead of treating patients.

Now suppose a third option, 'access only to computerized records', is introduced. A voter's fresh deliberations might change the ordering considerably. Perhaps access to electronically stored medical records would satisfy the demands of natural rights without impinging unduly on clinicians' time and so would be preferred to the 'no access' option. On the other hand, it might be claimed that access to only a subset of a person's medical records would throw up an incentive for doctors to exclude certain kinds of data from the computerized information system, thus circumventing any legal requirements pertaining to disclosure. Under the circumstances, it could be thought preferable to have access to all medical records, regardless of the form of storage. A person who thought that way would have preferences of the following nature:

---

[4] The problem is loosely based on comments of an MP made during a debate in the UK Parliament.

(2)   access to records > access to computerized records >
        no access to records.

This is not to say that the aspiring rational agent must have
the constellation of preferences implied by (1) and (2). But if a
person did have such preferences, for reasons such as those
outlined, it would hardly be fanciful to say that such preferences
were intelligible or coherent.[5] Rather, the difficulty lies in seeing
how aspersions of irrationality might be successfully cast. If
contraction is not a principle of rational choice, then this attempted logical defence transitivity fails. Claims to the contrary
fail because they ignore the logical distinction between binary
and ternary relations.[6]

INTRANSITIVITY AS INCOHERENCE: A CONSTITUTIONAL
CASE

Superficially weaker, though more appealing intuitively, a second
view holds that transitivity is embedded in the very meaning of
preference. An elegant expression of this 'constitutional' argument can be found in Davidson:

The theory ... is so powerful and simple, and so constitutive of
concepts assumed by further satisfactory theory ... that we must strain
to fit our findings, or interpretations, to the theory. If length is not
transitive, what does it mean to use a number to measure length at all?
We could find or invent an answer, but unless or until we do, we must
strive to interpret 'longer than' so that it comes out transitive. Similarly
for 'preferred to'. (Davidson 1980: 237)

Given developments in our knowledge both of behavioural
violations of SEU and of the mathematical theories consistent

---

[5] Note that it is an integral part of the example that the reversal is due to the
incentive compatibility concerns that are present in only one of the choice
situations. One might say, if this were an actual experimental result, that the
reversal was due to some kind of cognitive flaw. To raise epistemological issues
in this context would be to confound the thought experiment with thinking
about an experiment.
[6] Or rankings of any higher order.

with those violations, the methodological justification in the first half of the passage quoted, though appropriate at the time of writing, is not one we need to use any longer. However, the second part of the argument is deceptively simple and merits further consideration.

The defence is bound up in our view of measure theory in general, and in the properties of the relation 'is longer than' in particular. It would be difficult even to imagine how, with the possible exception of certain Wonderlands, a thing might have more than one numerical length. Even Alice's different heights were sequential: it was not as if she was two different heights at the same time. If length can be mapped on to the real number line, and the comparative 'is longer than' is the sort of relation that can be modelled by 'is greater than' (as normally defined over real numbers), then there is no escape from the conclusion that 'is longer than' will be transitive.

To admit this line of reasoning is not to forestall the questioning of its relevance. The crux of Davidson's argument is not the embeddedness of transitivity in the relation 'is longer than' but the similarity between preference and things being longer or shorter than other things, an analogy for which warrant must be provided. In fact, it could be said that the argument is a double analogy between the properties that relations have and the way both relations come to share a property. Both relations are transitive, and both have the property by virtue of the fact that it is embedded in their meaning. In neither case is transitivity a property the relation just happens, as a matter of brute fact, to have.

The modern practice is to avoid the use of analogies wherever possible. The main problem seems to be that there are no criteria by which any such arguments can, objectively, be judged valid. That they can be inconclusive is easily shown, and the existence of a 'counter'-analogy suggesting opposite results would provide such a demonstration. Perhaps the determination of a consumer's preference between two commodities, say, is more like the judgement of a competition in which players or teams compete in a pairwise fashion. Yet it is a feature of competitions that, while a first-division team might beat a second-division team that has, in turn, defeated a team from the division below, the third-division team could turn out to be giant-killers. This view holds that choice is the result of a cognitive competition and gives no grounds to presuppose transitivity.

There is no need to suggest that competition is a more plausible metaphor than length-measurement, though in some cases that may be the case. To counter the claim that rational preferences are transitive because preference can be likened to length, it is only necessary to point out that the existence of a transitive analogue is not sufficient to complete the argument. If it were adequate, preference would turn out to be an intransitive relation, as it can be likened to other relations for which transitivity does not hold.

THE MONEY PUMP

The money pump is a well popularized *reductio ad absurdum* which purports to show how an intransitive agent can be made to give up all his or her wealth. The argument is usually run on the following lines. Suppose an agent has an intransitive set of binary strict preferences as below:

(3)   *Pab*
(4)   *Pbc*
(5)   *Pca*

If these preferences can be translated directly into behaviour, then by (4) the agent will trade c for b and give up some strictly positive amount of wealth; by (3), the agent will trade b and some strictly positive amount of wealth for a; but by (5), a and yet more wealth would be given up for c. Such an agent is a dupe, and only a Panglossian fool would regard the cycle of trades as being for the best, so the story goes.

On the face of it, the argument is plausible enough. However, if the agent is not given complete information, in this case about the future sequence of opportunity sets, there is no reason to judge undesirable outcomes as the product of irrational choosing. It could be said that the losses suffered were the unsurprising consequences of having only partial information about future options. In addition, it might be noted that the temporal reading of transitivity implied by the money-pump argument cannot preclude the possibility that tastes might change over time, a prerogative that most decision-makers would not want to give up.

A mathematically inspired response to the above is to question time-bound interpretations and defences of transitivity. Properly, we should think of transitivity as a constraint that applies to preferences only at a single point in time, and it is this simultaneous interpretation from which the axiom derives its appeal.[7] We have now two interpretations, one dynamic and one simultaneous, both of which must be addressed.

Take the more appealing, simultaneous interpretation first. Useful as it is, the idea of a simultaneous interpretation is not entirely unproblematic. If it is taken to mean that the agent is confronted with three different feasible sets simultaneously, why do we not count them as one decision problem? Perhaps one could say that Pab just stands for the counterfactual proposition, 'if you could choose between a and b, you would choose a' and in the context of the money-pump argument, 'if you could choose between a and b, you would swap b and some strictly positive finite amount for a'. (3)–(5) would then become:

(6) $F = \{a, b\}$ $\square\!\rightarrow$ swap b for a and yield ε
(7) $F = \{b, c\}$ $\square\!\rightarrow$ swap c for b and yield ε'
(8) $F = \{a, c\}$ $\square\!\rightarrow$ swap a for c and yield ε''

where F is the feasible set, ε, ε', and ε'' are strictly greater than zero, and $\square\!\rightarrow$ is the counterfactual, '... if it were the case that ... then the following would happen ...'. The preferences in (6)–(8) say how a person would choose, given a variety of possible opportunity sets and assuming that the preferences are not transitive. To see how pernicious this is, we must follow the consequences of every possibility. Suppose the actual opportunity set turns out to be $\{a, b\}$: then, even with some strictly

---

[7] I am indebted to Frank Hahn for suggesting this line of reasoning.

positive transaction costs, the agent will trade b for a. Nothing to complain about in that. Now suppose that the feasible set were $\{a, c\}$. In this case, the agent would give up a for c and pay a positive sum for the pleasure. Again, there is nothing here on which to base a charge of irrationality, and the same conclusion would be drawn if we considered what would happen were b and c feasible. In so far as we can understand the constraint operating in a simultaneous sense, it seems that none of the possible outcomes has the disastrous consequences of which the money-pump parable is supposed to warn us.

Dynamic choice problems involve choice over time. In that context, there is a tendency to think that the problem with intransitive preferences lies with the consequences that befall the agent faced with a particular sequence of opportunity sets. It seems that, if a person is confronted with the choice sequence $\{a, b\}$, $\{b, c\}$, and $\{a, c\}$ where the feasible sets are offered at times t, $t + 1$ and $t + 2$ respectively, then they can be taken from a to a' and will pay $\varepsilon + \varepsilon' + \varepsilon''$ for the privilege. Formally, the concern is that, if we allow intransitive preferences, then a dynamic choice problem which is constructed by aggregating the antecedent conditions of the individual counterfactuals in (6)–(8) will prove to be the demise of any agent whose choices can be described by the aggregation of the consequents of those same counterfactuals.

As the song says, 'It ain't necessarily so'. Despite the similarities between conditionals and counterfactuals, there are differences of which we are becoming increasingly aware from the theories of David Lewis (1986) and Robert Stalnaker (1968). Here lies a difference that matters. In propositional logic, if it is the case that $P \rightarrow X$, $Q \rightarrow Y$, and $R \rightarrow Z$ (here upper-case letters denote propositions, and $\vdash$ is read 'proves that'), then $P \wedge Q \wedge R \vdash X \wedge Y \wedge Z$. The slogan might be: If you aggregate antecedents, you can construct the consequent of the aggregation by aggregating the consequents of the component conditionals. To see that this is an aspect on which conditionals and counterfactuals differ,[8] consider what might happen if, while

---

[8] The difference is due to the fact that the counterfactuals here are not constantly strict conditionals as material conditionals are (Lewis 1986: 31).

attending a conference in a foreign country, you were to lose some money. Suppose, in each of three possible variants of this event, that one of the things you would do is to drown your sorrows in a bottle of the local beer. The counterfactuals relating to the three forms of loss are:

(9)   lose cash □→ have a beer
(10)   lose credit cards □→ have a beer
(11)   lose travellers' cheques □→ have a beer

Whether your wallet with foreign exchange is stolen, the credit cards drop from your jacket, or your travellers' cheques are removed from your hotel room, you think that you would have a beer. However, there is also a small, ghastly chance that you could be deprived of all three forms of money simultaneously. From this, it does not follow that you would have three beers, because it might be that you would have nothing with which to purchase such solace as a beer might provide. Aggregation of consequents from individual counterfactuals does not follow from aggregation of antecedents as it does with conditionals.

Here is another version of the point. An experimenter asks a subject for preferences to three feasible sets and, on finding they are intransitive, points out that these preferences expose the subject to the dangers of being money-pumped. The subject then asks the experimenter on what grounds preferences over specific opportunity sets are used to infer behaviour over the concatenation of those sets. The subject adds that, if the experimenter had asked what choices would be made in the super-game now proposed, his choices would be different. Moreover, given the intransitive preferences he has, it is absurd to use them as the basis for a decision rule in the dynamic choice problem now under consideration. Their use would indeed be irrational, but who was it that was proposing the application? What I take to be the same point is crisply put elsewhere. Irving LaValle and Peter Fishburn write:

Sensible people with cyclic preferences would simply refuse to get involved in the money pump game unless they were deceived into believing it was to their advantage to do so. (LaValle and Fishburn 1988: 1225)

DECIDABILITY

A criterion that drives many theoretical developments in economics and game theory derives from the concern that theories of decision-making should make predictions that are uniquely determined by the description of the decision-problem. Theories that have this property are sometimes said to be decidable, and those that permit intransitive preference relations are claimed to be not decidable. If a person has intransitive binary preferences over three choice objects—a, b, and c, say—then no prediction can be made about choice from the feasible set $\{a, b, c\}$.

The argument is not unlike the attempted proof by contradiction used in the attempted logical justification of transitivity discussed earlier. Suppose a decision-maker with intransitive binary preferences in (3)–(5) has the preference ranking *Tabc*. This is a precise, unique prediction about the decision-maker's preferences. It satisfies the requirement of decidability and yet permits a violation of transitivity between the binary preferences. Indeed, if the decision-maker had given the preference ranking *Pcba*, or any other of the eight total possible orderings, these too would have been decidable. In short, it is completeness, rather than transitivity, that determines whether preferences are decidable.

This defence of intransitivity does not come free, as it removes the co-relations between preferences which an external observer might use to predict choices in other contexts. Without such co-relations, the external observer can say little about a person's behaviour based on sample information about that person's behaviour. For sure, this raises the costs of predicting behaviour, but why should that be a concern for the decision-maker herself? That we should be able to make certain inferences from limited observations is obviously important for positive theories, but it is difficult to see why it should be of concern to an agent selecting a rule by which to act.

RATIFIABILITY

It is sometimes said that the good decision is one you would not want to change once you have made it. Decisions that are the cause of instant regret are said not to be ratifiable. *A fortiori*, they are not taken to be rational, and it is claimed that intransitive preferences can lead to unratifiable decisions.[9]

Suppose you have to choose from the set $\{a, b, c\}$ and that you possess the intransitive preference structure as before. Then for any particular choice there will always be something you would rather have chosen: the selection of a is not ratifiable because c is preferred to a, and similar reasoning can be applied to undermine choice of b or c. However, if rational choices must be ratifiable and intransitivity makes all choices unratifiable, then no rational choice is open to the person without transitive preferences.

The problem with this argument is similar to that used against the proof by contradiction. If one thinks that rational choices need to be ratifiable though not transitive, one merely has to point out that there is no logical reason why a person choosing from three items should see their binary preferences as relevant. No choice out of three items would be ratifiable, either, if the choice were not in the set of maximally preferred items or if the subject had contradictory preferences (violating completeness). Neither of these possibilities follows from the fact that the binary preferences are intransitive.

COUNTER-EXAMPLES

The balance of the discussion so far has been to focus on the logic of arguments for the normative justification of transitive preference. The case for intransitive preferences is usually motivated by a subjectivist desire to be thoroughly Humean or to adopt a 'hands-off' approach to preferences, whereas counter-examples tend to have greatest persuasive force when there is

---

[9] I am grateful to Bill Harper for making me aware of this line of argument.

some kind of quasi-objective view about what should be done. There is no reason to expect that a full-blooded subjectivist theory will find any persuasive counter-examples. These may be the least persuasive aspect of the intransitivity literature, but that is not to say that they have no merit or should not be mentioned in discussions of this kind. Some of the examples that appear to support the rationality of violating transitivity will be discussed under the following classes: context dependence, multiple criteria decision-making, and sensitivity to counter-factuals.[10]

Context Dependence

You are at a friend's dinner party and your host is about to offer you some fruit.[11] If you are offered an orange or a small apple, you'd rather have the orange, and if the choice is between a large apple and an orange you think you'd rather have the large apple. As it happens, your friend runs out of oranges and emerges from the kitchen with two more apples, one large and one small. How should you choose?

Perhaps there are parts of the world where you should select the large apple just because it is the biggest. Indeed, transitivity dictates that you pick the large apple. But if you want to pick the small apple, must you be condemned as irrational? This is not to say that you must select the small apple, but it is just the kind of choice that follows from the inculcation of manners in young children. Furthermore, it is difficult to believe that any parent or guardian would be persuaded that the polite choice (small apple over large apple) is the wrong choice, even when it is intransitive.[12]

---

[10] An extensive examination of plausible counter-examples appears in Bar-Hillel and Margalit (1988) to which interested readers are referred.

[11] I thank Bob Sugden for this example, though its original provenance is unknown.

[12] Implicit in this example are the assumptions that agents can act on the basis of non-utility considerations and that politeness is one such consideration. Before the mathematics of non-transitive choice was as well documented, the reaction to such examples was to say that, if the intransitivity was compelling, there must be a reason for it. If a reason exists, then it should be reflected in the

Table 4.1

Options	States					
	s1	s2	s3	s4	s5	s6
α	1	1	4	4	4	4
β	3	3	3	3	3	3
γ	5	5	2	2	2	2

The following example takes the discussion from decisions to games. The statistical 'paradox', formulated by Blythe (1972) and discussed elsewhere by Packard (1982), shows that choice over pairs of lotteries can be intransitive for players who want to maximize the probability of winning. To see this, suppose there are three options, α, β, and γ, with consequences depending on the outcome of six possible mutually exclusive states (s1–s6), as in Table 4.1. A player is offered the opportunity to choose one die from each pair. The dice are fair and highest score wins. The opponent throws the other of the two dice and each player wants to maximize the probability of winning. How should a player choose?

Faced with a choice between α and β, the probability of winning is $\frac{2}{3}$ with α and (therefore) $\frac{1}{3}$ with β, which makes α the option that maximizes the agent's chance of winning. Comparing β and γ, it is apparent that γ is dominated by β, which, of the two, gives the highest probability of winning. But now consider the comparison between α and γ. To begin, it is not possible to win with α if 1 comes up. There is a $\frac{2}{3}$ chance that α will show at 4, but even if it does there is a $\frac{1}{3}$ chance that γ will beat it with a 5. The only way to win with α is if the outcomes are 4 and 2 and, as this occurs with probability $\frac{4}{9}$ ($\frac{2}{3} \times \frac{2}{3}$), the maximizing choice is γ.

---

descriptions of the options, in which case 'a small apple when the alternative is a large one' is not the same as 'a small apple when the alternative is an orange'. This makes the behaviour consistent with transitivity, but then, as there is no end to its application, all behaviour becomes transitive and the axiom ceases to become testable, contrary to assumption. This is proved in Ch. 6.

Multiple Criteria and Incommensurability

Incommensurability is thought to provide an explanation for intransitive choices, and the idea is consistent with a common view that action is, in part, an attempt to construct meaning within the world. As usually described, problems of commensurability normally arise when a choice has to be made between options that differ with respect to the attributes they possess. For instance, suppose a development agency were to follow the rule, 'select that project which compares most favourably on the criteria shared by all options being compared', and that the options would offer the following benefits:

- *Option a:* produces 200 litres of clean water per day and prevents 100 cases of typhoid a year.
- *Option b:* prevents 170 cases of typhoid a year and increases village production by 120 bushels of maize a year.
- *Option c:* increases village production by 150 bushels of maize per annum and produces 80 litres of clean water per day.

It is easy to see that, following the rule above, the agency will prefer *b* to *a* and *c* to *b* while preferring *a* to *c*. In the instances where decisions have to be justified to others, the desire to make comparability preferring choices is understandable. This seems to happen extensively in the public sector and reflects the fact that decisions both influence and determine the social construction of what it is meaningful to do.

Counterfactual sensitivity

Agents sensitive to the counterfactual are concerned not only for what they do, but for what might have been or what they might have done. The sentiment of guilt would not, for instance, be intelligible if the agent could not have done something else. More prosaically, suppose you wanted to make a choice that took account of what would happen both if you chose an action and if you rejected it—formally, we might say that you had a bivariate utility function in the form $u\{$accept option i, reject option $j\}$. Further, imagine that the choice is between three options with acceptance and rejection primes as below:

Options	Accept	Reject
One	5	3
Two	3	$\sqrt{2}$
Three	1	1

Now suppose a decision-maker wanted to choose the option that maximized the value of a particular bivariate function, say the following:[13]

$$u = accept_i + (reject_j/accept_i)^2.$$

It is a matter of straightforward calculation to see that the agent would have intransitive preferences over the three preferences, and there are a number of well developed mathematical theories which make the point more generally.[14]

Though opinions are shifting, there is no consensus as yet on whether regret is rational. In one sense it is not. If a person had certain beliefs and desires that point to an act being optimal *ex ante*, then he should do that act. It may not, *ex post*, turn out to have been the best act, and that might lead the person to suffer what might be called regret. In this case, if sensitivity to the expectation of regret causes the decision-maker to pick a different act, it is one that is, *ex ante*, dominated by the optimal act. Such behaviour could not be rational.

Regret could, on the other hand, be taken as something that determines the preferences themselves. For instance, Douglas and Isherwood (1980) take a Veblenite view of consumption and maintain that the acquisition of particular goods can serve the function of signalling the social groups to which a person wants to belong.[15] Not going to church on Sunday is still a very clear rejection of social norms in rural Ireland. Similarly, not owning a car to park on the drive is for some an expression of their concern for the environment. Sensitivity to what might have

[13] I am grateful to G. T. Jones and Bob Sugden for their comments on an earlier rule.

[14] See e.g. Morrison (1962), Krantz *et al.* (1971), Bell (1982), Fishburn (1982), and Loomes and Sugden (1982).

[15] These are known in psychological parlance as 'reference groups', and are to be distinguished from the groups of which a person is actually a member.

otherwise happened is also strong in legal settings, where not performing certain things can lead to charges of negligence. Where organizations have to take account of the *ex post* benchmarks that the media and others use to evaluate their activities, there are strong incentives to consider what others may say might have been. In short, whether regret is an aspect of the rational or irrational depends on the sense in which it is being used.

CODA

The focus of this chapter has been the logic of arguments for adhering to the transitivity constraint on preferences. In addition, some attention was given to counter-examples in which violation of the axiom seems reasonable. Although the judgement that violation of transitivity can be reasonable is of substantial interest in its own right, it is germane also to our acceptance of the new mathematical theories that generalize the idea of utility. *If* the case put is accepted, then we cannot dismiss such theories as accounts of irrational choice. In any case, the reading of the literature I prefer is that, if a logically valid defence of the normative interpretation exists, we do not yet know what it is. Chapters 7 and 8 give positive reasons to suggest that such arguments do not, and could not, exist.

5

Independence

INTRODUCTION

Rational choice theory knows of many independence axioms. Because they indicate what aspects of individual or social choice can be treated independently, they are key to the practice of understanding problems by decomposing them into smaller units. Mathematical expressions of different independence assumptions bear considerable similarity, although this should not be allowed to mask the fact that their meanings, and plausibility, differ considerably. As noted, SEU uses independence in a sense that amounts to requiring that utility is linear with respect to probability. The value of a gamble, therefore, is the probability-weighted sum of its component pay-offs, which we can write $\Sigma p \, u(x)$.

Until the end of the 1980s, some theorists held the view that the independence axiom was, of all SEU's postulates, the most important.[1] Unlike transitivity, the linearity assumption is explicitly of crucial importance in statistics, where, roughly, it appears as the expectation operator. There is reason, then, to suppose that, if a normative justification for conventional utility theory exists, the strongest arguments will be those made in defence of linearity. However, we saw in Chapter 2 that agents consistently violate the assumption, and that there are now many mathematical models that capture such behaviour. Are such violations normatively acceptable? It would seem that there is now a consensus, not about the normative status of the assumption, but about the methodological decision that its interpretation

---

[1] '[i]t is the independence axiom which gives the theory its empirical content . . .' (Machina 1982).

necessitates. This chapter outlines the main arguments leading up to the current position.[2]

COMPLEMENTARITY

Early discussions of independence as linearity read as if there is barely anything to discuss. The 'absence-of-complementarity' argument first appears in the von Neumann and Morgenstern (1944) as the brief remark that 'the two alternatives are mutually exclusive so no possibility of complementarity and the like exists'. Given that lotteries are defined on sets of mutually exclusive states, only one consequence will ensue. As a result, there can be no question of the decision-maker enjoying more than one consequence, nor does the possibility of complementarity between multiple consequences arise. Samuelson (1952) takes a similar tack, though he notes that independence is not tenable in the context of choice between ordinary commodities where complementarities are rife.

If we take this to be an argument about the impossibility of contemporaneous consequences, then we have no choice but to admit that the conclusion follows validly from the definition of a lottery. However, as McClennen (1982) points out, the defence is weak in the sense that pointing to the absence of complementarity only removes a possible explanation for the existence of non-linear preferences; it does not follow from this that no such argument exists. Viewed *ex ante*, then, a lottery is a set of chances of pay-offs that are enjoyed simultaneously. To be sure, this does not compel anyone to violate independence, but it shows that such violations can be explained and cannot, therefore, be ruled irrational on the grounds that they are unrationalizable.

COUNTER-EXAMPLES

Even if behaviour is explainable, this may not necessarily mean

---

[2] In doing so, and without claiming to do justice to its rigorous detail, I draw particularly on work by Ned McClennen.

that we must call it rational. (In behavioural or subjective theories the difference between 'being able to explain' and 'judging rational' may, however, be very small.) To help judge the rationality of any rule violation, it is helpful to examine the details of specific cases.

The first and now celebrated counter-example of independence was produced by Allais (see Chapter 2). Recall that, in informal experiments, Allais asked subjects how they would choose in situations 1 and 2 (see p. 23). The assumption of independence at the very least requires that strict preference for A/B should be matched by strict preference for A'/B' although the modal constellation of choices was B and A', a finding replicated in a number of subsequent experiments. From a standard economic viewpoint, the evidence may seem doubly inexplicable, for economists have long since allowed that agents with higher wealth levels can be expected to be more risk-preferring. The expected wealth of the chooser is considerably higher in situation 1, but subjects shift to a riskier option in situation 2.

For all that, one does not have to possess super-human powers of imagination to see how subjects might argue intelligibly for B and A. The certain prospect of £1 million is sufficiently attractive that it should not be given up for option A, which brings with it a 0.01 chance of winning nothing. On the other hand, both options A' and B' offer only small chances of winning a prize, around 0.1. However, A' offers £5 million instead of £1 million if the chooser wins, so why not pick A'?

Like Allais, Machina (1981) proposes an example in which violation of independence appears appropriate. However, he uses a decision problem in which the consequences are non-monetary pay-offs and a line of attack which is aimed at providing positive support for a complementarity argument. The problem is this. A friend is in hospital about to undergo surgery to his leg, and you have to decide whether to send flowers and a sympathy card ('Flowers') or champagne and chocolates ('Champagne'). The consequences about which you are ultimately concerned are fully described thus:

α: leaves with permanent limp, and receives flowers and sympathy card from you

β: leave with permanent limp, and receives champagne and chocolates from you

γ: comes out in perfect health, and receives either flowers and sympathy or champagne and chocolates from you

δ: dies on the operating table, and doesn't see either the flowers or the champagne you have sent

Two possible choice scenarios leading to these consequences are described in Table 5.1. Suppose you find yourself in scenario A.

Table 5.1 The gift problem

	Outcome	
Action	(limp) = 0.05	(no limp) = 0.95
Scenario A		
Flowers	α	γ
Champagne	β	γ
Scenario B		
Flowers	α	δ
Champagne	β	δ

You may think that, given the substantial probability of your friend recovering, 'Flowers' would be appropriate, just in case your friend ended up with a limp. Were you then to consider a second scenario in which the most likely event entailed your friend's death, you might think that 'Champagne' is appropriate, in case he survived. While it is possible to understand why a person might have the preferences that such a pattern of intended choices implies, it is evident that such a pattern constitutes a violation of independence. However, if preference depends on what might have been (death in one scenario, complete recovery in another), then it can surely not be irrational to take this into account before the fact.

Any supporter of the subjectivist position who regards argument about preferences as beyond the pale is bound to accept Machina's version. However, the force of the example has been questioned on the grounds that the preferences themselves are not as intelligible as we might think (Sobel 1988). Why, given the description of the consequences as they stand, should a decision-maker have any strict preferences at all, let alone those proposed? Recall that the consequences that define (α)–(δ) must be complete descriptions of all that matters to the decider, and that everything without those descriptions is excluded if, and only if, it is immaterial. Among those things not included in the formal descriptions of the consequences is the reaction of your friend upon receipt of your gift, though this would seem to play an important role in the story as it stands. Sobel allows that we might allow a non-consequentialist justification of independence violation in terms of doing the right thing, but inclines more to a modification of the paradox, which adds information about the patient's response to your action. These new consequences are:

α' : α *and* my friend smiles at my offering, and says 'He cares. What a pal!'

β' : β *and* my friend groans and says 'What is this, some kind of sick joke?'

α'' : α *and* my friend winces and says 'What does this mean? What's the guy trying to tell me?'

β'' : β *and* my friend, bursting with joy, says to no one in particular, 'Hey let's party!' and does.

This variant makes explicit factors which not only warrant attitudes other than indifference but also rationalize the constellation of preferences. Sobel suggests that this kind of objection is best regarded as an objection not just to the independence axiom, but to utility theory *in toto*. Nevertheless, objection it is.

DOMINANCE

To violate linearity is, it is claimed, to violate first-order stochastic dominance.[3] A statement of the 'proof' can be found in Weber and Camerer:

[3] Strong first-order stochastic dominance implies the following condition: for all possible outcomes x_i and y_i from actions x and y, if $\Sigma_{x_{min}}^{x_i} p \leqslant \Sigma_{y_{min}}^{y_i} p$ with strict

these [non-EU] theories have one unattractive quality. . . . The preference function for each component need not be linear in probabilities
. . . but non-linearity leads necessarily to violations of stochastic dominance. Excluding violations of dominance implies [linearity]. (Weber
and Camerer 1987: 134)

The mathematical proof that accompanies this text is simple and
elegant, but it is not evident that it furnishes the proof one
might expect. At the end of the proof, Weber and Camerer note
that violations of stochastic dominance preference are certainly
undesirable, though they do not say why. For practical purposes,
such things are often taken as given, but when it comes to the
normative evaluation of decision rules, it is precisely this kind of
assumption that we must examine. The lesson to be drawn is
that, while we might accept the mathematical component of a
normative claim (in this case, the link between linearity and
first-order stochastic dominance), we are by no means bound to
accept claims about its significance.

More interesting, perhaps, is the fact that even the proof itself
raises some curious methodological questions. To see these,
consider the structure of the Weber–Camerer version. First,
non-linearity is characterized as follows:

$$u(\text{pay-off} = \pi, prob = p + q) >^{[4]} \tag{1}$$
$$u(\text{pay-off} = \pi, prob = p) + u(\text{pay-off} = \pi, prob = q),$$

where $u(\cdot,\cdot)$ is a utility functional defined over pay-offs and their
probabilities (*prob*). Two lotteries are then defined:

$$L1 \equiv \{(\pi, p + q), (0, 1 - p - q)\}$$
$$L2 \equiv \{(\pi + \delta, p), (\pi, q), (0, 1 - p - q)\}.$$

Note that $L2$ first-order stochastically dominates (FOSD) $L1$.
The proof is completed by noting that if (1) is true then there is

inequality holding at least once, then Pxy. X_i are assumed to be ordered with
respect to preference, and x_{min} is the least preferred pay-off from $x$ing.

[4] Alternatively $<$, but we ignore this just to minimize the notation.

a δ sufficiently small such that:

$$u(\pi, p + q) + u(0, 1 - p - q) > u(\pi + \delta, p) + u(\pi, q) + \qquad (2)$$
$$u(0, 1 - p - q).$$

The inequality in (2) describes an ordering that is the reverse of that derived by application of the FOSD rule. The proof uses nothing much more than definitions of lotteries, FOSD, and the assumption that utility is linear in the probabilities (LP), but the step in which δ, the small increase in the pay-off, is introduced depends crucially on the assumption that utility is continuous in its first argument (CON). This means that the proof shows that violation of LP *and* adherence to continuity lead necessarily to violations of first-order stochastic dominance. It says nothing about agents who violate both LP and CON.

There is more. Consider another implication of non-linearity and a special case of (1) in which $p = 1$ and $q = 0$, from which we can derive the following:

$$u(\pi, 1) > u(\pi, 1) + u(\pi, 0). \qquad (3)$$

Because the last component has probability zero, this can be ignored and the remaining terms can be used to define trivial lotteries in which the pay-off is π. This results in a contradictory proposition:

$$u\{(\pi, 1)\} > u\{(\pi, 1)\}. \qquad (4)$$

If the CON assumption is applied, then the following can be added:

$$u\{(\pi - \delta, 1)\} > u\{(\pi, 1)\}, \qquad (5)$$

which is one step more bizarre than (4) and apparently represents a violation of dominance. If (5) is read as showing that a dominated action is preferred, then it might seem that this is a stronger argument for LP, as it needs no recourse to the (at least slightly less safe) notion of stochastic dominance. Alternatively, it could be claimed that, because δ in (5) is a positive increment in the consequences, as assumed before, the assumptions and procedures used to get from (3) to (5) cannot be consistent. As contradictions are consistent with any proposition, it should not be surprising, or worrying, if they can be shown to be inconsistent with FOSD.

To see a final, and even more transparent, paradox that this proof type throws up, consider the following inequality:

$$u\{(2\pi, 1)\} < u\{(\pi, 1)\} + u\{(\pi, 1)\}. \tag{6}$$

Interpret this as saying either that the marginal utility of money declines or that the agent is risk-averse. Now, by CON, derive

$$u\{(2\pi + \delta, 1)\} < u\{(\pi, 1)\} + u\{(\pi, 1)\}. \tag{7}$$

If this represents a violation of dominance, which it seems to, then presumably we should conclude that either risk aversion or declining marginal utilities of money are irrational (depending on the interpretation chosen). Neither seems a helpful conclusion to draw.

An attempt might be made to argue why the use of the continuity assumption is appropriate in the first proof and not in the paradoxical proof above, though I know of no such argument. It might be claimed that (7) does not offend the decision-theoretic sensibilities in the same way that (2) does, but again we might find it hard to see why these types of behaviour do not fall or stand together. Rather, it seems that using continuity is less innocuous than one might suppose. The proof might just as well be turned on its head. For those who accept that non-linearity is normatively acceptable, this could be a foil against FOSD. Perhaps FOSD is not as fundamental as it seems; or perhaps it is only risk neutrality that is rational. Holders of an intuitionist philosophy of mathematics can certainly make sense out of this. Continuity is usefully applied when deriving differentials but makes little sense here, so it should not be used.[5]

---

[5] A somewhat related methodological point is made in Rubenstein (1989), where he shows that an assumption of continuity cannot be relied on to extend analysis of finite games to games with no known termination. A discussion in Machina (1989) takes a slightly different view but makes the point that some non-expected utility theories can be consistent with FOSD under certain 'monotonicity' conditions. The Samuelson–McClennen (Samuelson 1952, McClennen 1988) line is that the unacceptable-implications argument (FOSD) depends on the application of just another version of independence. Broadly, I think the argument presented here is in line with their view.

DYNAMIC CHOICE

Thus far, the arguments concerning linearity have presupposed the canonical one-shot decision problem in which the agent selects an act and nature picks a state. There are many situations, however, in which decision-making may take place after uncertainty has been partially resolved. One of the more interesting lines of defence for linearity exploits the framework that these 'dynamic choice' problems offer and attempts to show that violations cannot be justified and may lead to undesirable consequences.

A particularly clear introduction to this line of reasoning can be found in Raiffa (1968), who proposes a problem on which the following is based. Imagine that you are a contestant in a quiz show and that at 7.00 p.m. you will be given a chance to spin a wheel of fortune. Of the wheel's 100 divisions, 89 are painted red. If the wheel stops spinning at any one of these red divisions, you will win a mystery prize. However, if the wheel comes to rest at one of the remaining 11 green divisions, then you will be given an opportunity to take either £1 million in cash or a 91:9 chance of winning £5 million or nothing. The decision tree for the problem is shown in Fig. 4.

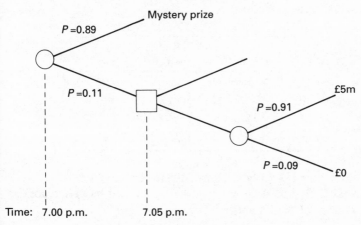

Fig. 4 A dynamic choice problem

Now consider the following questions that Raiffa poses:

- *Question 1:* If the wheel of fortune were to land on one of the green divisions, would your choice between the £1 million and the 91:9 lottery depend on a 'detailed description' of the mystery prize?
- *Question 2:* If the show's host were to ask you before the show how you would choose if you were not to win the mystery prize, would your decision depend on what the mystery prize was?
- *Question 3:* If, after the spin, you were in the position of having to choose between the £1 million and the 91:9 lottery, would the choice differ from that declared before the show?

Raiffa's own view was that one should give negative responses to each question, and he showed how such replies could be construed as being consistent with the linearity assumption. If your choice between the sure pay-off and the lottery at 7.05 p.m. is independent of what the mystery prize would have been, then there is no reason why you should postpone declaring your preference at 6.55 p.m., and that declaration would match your choice at 7.05 p.m. If the mystery prize were £1 million, then the choice would be equivalent to a choice over actions $A1$ and $A2$ as described in Table 5.2. If, on the other hand, the mystery

Table 5.2

Actions	$p = 0.89$	Outcomes $p = 0.10$	$p = 0.01$
$A1$	Mystery: Q	£1m	£1m
$A2$	Mystery: Q	£5m	£0
$B1$	Booby	£1m	£1m
$B2$	Booby	£5m	£0

prize turns out to be a booby, then the preference for the sure pay-off over the lottery would be represented by preference for $B1$ over $B2$. Raiffa writes:

In any event, if you agree with the negative answers to Questions 1 and

2, the logic seems to indicate overwhelmingly that you should not persist in choosing [*A*1 over *A*2 and *A*3 over *A*4]. (Raiffa 1968: 83)

A particularly careful analysis by McClennen (1982, 1988) indicates the value of examining the argument in its entirety. Broadly speaking, there are three issues worth further consideration: the strategy of the proof, the link between the dynamic choice problem and violations of linearity, and answers to three questions. We deal with these in reverse order.

Suppose that the wheel turns, and you miss the mystery prize which you are told was a £3 million yacht. To yourself, you might say that, having lost once, you should play safe and take the £1 million. In another possible world, you learn that the prize was a booby—a year's supply of pickled cabbage. You can't do any worse than you have done, so why not take a gamble? In both worlds, the dynamic consistency defence of SEU judges you to be irrational—not because of the choices you make in any actual world, but because of the relations between the choice you make and events that might have occurred in possible worlds that happen not to have occurred. From a subjectivist standpoint, the onus would seem to be on those who claim that such reasons are normatively insufficient to say why such behaviour is irrational.

The link between the dynamic choice and violations of linearity seems no more conclusive. In transferring from dynamic to static situations, information is, of necessity, thrown away. Defenders of independence may argue that the parts of the dynamic problem that the static description strips out are of no informational value. Conversely, it is possible to object to the reduction on the grounds that it removes precisely the information that distinguishes static and dynamic choices. Negative replies to each of the questions above, therefore, need not have any relation to choice in the static problem.

The third issue concerns the strategy of the questions as a form of proof. The questions seem general enough, appealing as they do to the independence of choice from the peculiarities of the mystery prize, whatever it is. However, in so far as we are asked to consider only one social situation, i.e. a game show (and an experiment in the original version), the test is actually very narrow. This seems to be a non-trivial methodological issue. Even if one accepts the negative answers to the three questions and the equation of static and dynamic choice problems, there remains the question as to whether independence appeals in the wider class of economic and social situations. Advocates of independence might argue that we best understand principles that underpin rational choosing first in the abstract. Opponents will say that these quasi-abstract formulations rob problems of just the kind of material that makes more general forms of behaviour seem attractive.

Technical analysis related to the questions posed can be found in work by Peter Hammond (1976, 1988a). Roughly, Hammond emphasizes the importance of final consequences—a doctrine that he calls *consequentialism*[6]—and examines the implications of taking only these into account in dynamic-choice problems. He proposes that the set of preferred acts be the same at the outset of any decision tree, in the planning stage, and at the point where the decision must be made. Using this notion of consistency as an assumption of consequentialism, it is then possible to show that linearity follows (Hammond 1988b). The result is intriguing in that it seems to be constrained only by the requirement that events with zero probability must be excluded from the decision tree used to map the problem. What more substantive grounding could one ask for any assumption than that it requires the agent accounts for all, and only all, of the consequences of his actions?

Work by Ned McClennen (1986) suggests that the conceptual debate is less clear-cut than the theorems of consequentialism might at first suggest. Suppose a decision-maker is sophisticated in the sense that she takes into account preference changes at the planning stage (Strotz 1956). Such a person might well intend to take $A1$ in preference to $A2$ and yet point out that, when the nature of Q is known, neither $A1$ nor $A2$ is feasible. Selection of either $B1$ or $B2$ will, for a person who has good

[6] This is not consequentialism in the formal, philosophical sense discussed in Scheffler (1988).

reason to stick to her plans, be independent of her choices when $A1$ and $\bar{A}1$ are both feasible.

McClennen also devotes considerable attention to what he calls the *resolute* decision-maker, a person who, in effect, uses the fact of his own planning to bind himself to courses of action that he would otherwise reject. If plans take into account what has gone before, then it is perfectly possible to make different choices at 7.05 p.m. depending on whether the decision is *de novo* or merely part of a series of decisions for which a plan was made and is now being followed. The question is whether plans should be made in such a way that they are sensitive to what has gone before. For one thing, resolutions of uncertainty provide new information, which we expect (in general) to make changes in behaviour rationalizable. However, one could argue that the decision-maker plans for choice only on the assumption that a choice may have to be made, and that whenever it does the assumption is vindicated. *Ergo* the assumption is innocuous.

The last word (almost) in this debate goes to Machina's (1989) critique of normative arguments for linearity. Whether decisions can be sensitive to choices taken or risks borne prior to the current problem is a query that recurs both in debates directly relating to consequentialism and in other attempted defences of linearity. Proposing a test of intuitions in only one social context might be questionable, and the following problem proposed by Machina seems to confirm this. Two children have been promised an outing, and they have also been promised that one of them can choose it. Their father says he prefers that they toss to determine who should make the choice. A coin is duly tossed, and when it lands heads up the child who loses says: 'Ah, but you say that you would strictly prefer that we toss the coin', and she will go on saying this until she wins the call.

In effect, the child is asking the parent to act as if they had not been through earlier parts of the decision tree. However, one can draw an analogy with the standard economic treatment of non-separable preferences over time for non-stochastic goods. Whether you prefer melon to cheese may depend on whether you have had just meat or fish, and similarly with lotteries. Whether you prefer a sure £1 million may well depend on

whether you have already borne the risk of taking the consequence Q so that, formally, your preferences before 7.00 p.m.
should be stated in terms that make this conditionality explicit.
We pay insurance company premia not for a pay-off, but in
exchange for its bearing our risks, which requires only that a
contingent payment be made (Anand 1987). We do not ask for the
premium to be returned if the contingency does not arise.

We might say that a decision-maker who resolves to choose
the lottery prior to the knowledge of Q but chooses the certain
£1 million at 7.05 p.m. is a person for whom the following
inequalities hold: $v(Q$ gone, lottery) $<$ $v(Q$ gone, £1m); and $v(Q$
possible, £1m) $<$ $v(Q$ possible, lottery). This is an example of
what Sen (1987: 61) calls a 'refined description', and it shows
that what is, apparently, a species of preference reversal could
nevertheless be part of an agent's plan. To evaluate courses of
action at different points in the decision tree without reference
to where one is, and where one might have otherwise been, is to
employ a consequentialist decision rule which is, in general,
inappropriate for agents with non-expected preferences. To do
so leads to the prescription of counter-preferential behaviour, so
it is no wonder that such devices can be used to make agents
look silly.

The sting in the tail of Machina's review of linearity is that he
does not conclude against the assumption. Many writers have
questioned whether concern for what might otherwise have been
(sensitivity to the counter-factual) cannot be captured within the
definition of a consequence. John Broom (1991) argues that it
can, and that all preference-related issues can be taken into
account when individuating consequences. In doing so, he reminds us of a view that can be found, if not followed through,
in Savage (1954: 8–16). Machina, we noted, offers a compromise
that keeps the normative status of linearity at the deepest level
of description while allowing that, at the level of description
with which economists work, linearity cannot be expected, even
of rational agents. Whether this final proposal is accepted has
yet to be seen, but I suspect that the parent-and-children example
is too convincing for the deal to be plausible, and that the deal
itself will be too pragmatic for others.

6

Completeness and Continuity

INTRODUCTION

In this chapter, we examine the normative appeal of two more axioms: completeness and continuity. Precise definitions will be given later, but informally, completeness requires the agent to have a preference ordering for every possible choice set, while continuity excludes the possibility of consequences so extreme that they matter however unlikely they are. Neither of these axioms, it must be said, has received much normative support even from advocates of SEU. Some defenders of SEU have tried to dismiss the problems that they raise on the grounds that the assumptions are not 'substantive'. Perhaps there is something in this, though just saying it hardly establishes a case. Against that, there are discussions in the literature in which, as we shall see, the assumptions quite clearly have non-trivial behavioural implications.[1] Some of the arguments for continuity verge on the eccentric. Those for completeness are hardly more compelling, although the assumption provides a better foil for discussing the variety of interpretations to which formal axioms are prone. For any good theory, formal constraints on behaviour should be acceptable at all the points of behavioural interpretation.

Completeness

As a preliminary, it is useful to recall that work by a number of mathematical economists, including Mas-Collel (1974), Gale and Mas-Collel (1974), Shafer (1976), and Kim and Richter (1986),

---

[1] We allow for the moment that claims about behavioural content can be substantiated.

shows that many of the theorems of general equilibrium and consumer theory can be reproduced without resort to the assumption that preferences are complete. As with the axioms previously discussed, the theoretical issues to be resolved are, therefore, not technical but rather normative. Based on the version introduced in von Neumann and Morgenstern (1944: 26), we may formally define completeness as follows:

$$(a)(b)(Pab \text{ or } Pba \text{ or } Iab). \qquad (1)$$

(1) can be read: 'for all a and b, either a is strictly preferred to b or b is strictly preferred to a or a and b are equally attractive'; we use 'or' instead of ' $\vee$ ' to exclude the possibility that both elements of the disjunct are true. In the original statement of the axiom the symbols ' $>$ ' and ' $=$ ' are used instead of P and I, and while it is possible to read the notation in a purely formal spirit, predicate notation is less leading about the properties that relations might possess. Von Neumann and Morgenstern consider the 'intuitive meaning' or 'justification' of the axiom and point to the fact that a similar assumption is made by indifference curve analysis. But suppose the justification for using completeness in indifference curve analysis were *ad hoc*: do two wrongs make a right? They write:

It is conceivable—and may even in a way be more realistic—to allow for cases where the individual is neither able to state which of two alternatives he prefers nor that they are equally desirable. . . . How real this possibility is, both for individuals and for organisations, seems to be an extremely interesting question, but it is a question of fact. (von Neumann and Morgenstern 1944: 19)

Savage's (1954) definition is more economical. A single primitive weak preference relation, R, read as 'is at least preferred to', is used to define completeness thus:

$$(a)(b)(Rab \vee Rba). \qquad (2)$$

(2) can be read: 'for all a and b, either a is weakly preferred to b or b is weakly preferred to a'. Respectively, indifference and preference can be derived from the conjuncts $(Rab \wedge Rba)$, and $(Rab \wedge \neg Rba)$ or $(Rba \wedge \neg Rab)$. Savage makes clear that he regards the axioms as normative constructs and draws an

analogy with logic, claiming that axioms should be used to police decisions for consistency. Towards the end of the relevant section of *The Foundations of Statistics*, he notes:

This [absence of completeness] would seem to give expression to introspective sensations of indecision or vacillation, which we may be reluctant to identify with indifference. . . . My own conjecture is that it would prove a blind alley, losing much in power and advancing little, if at all in realism; but only an enthusiastic exploration could shed light on the question. (Savage 1954: 21)

Rather than persuading us that it is irrational to violate the completeness axiom, this passage might leave readers asking: What is meant by power (e.g. is it explanatory, predictive, or something else?), and why should a decision-maker, as opposed to an observer, be concerned about it? Furthermore, if we accept Savage's claim that the axioms are normative, logic-like constructs, then we are not in a strong position to understand the relevance of his remarks about sensations of indecision and vacillation—presumably, observations about states of mind that we have, as a matter of mere fact.

Aumann is less enigmatic and states simply:

Of all the axioms of the utility theory, the completeness axiom is perhaps the most questionable. Like others it is inaccurate as a description of real life; but unlike them we find it hard to accept even from the normative viewpoint. (Aumann 1962: 446)

Viewers of *Call My Bluff*[2] may feel a certain sense of *déjà vu*.

Is completeness an interesting question of fact, an inaccurate description of real life, or an expression of indecision and vacillation? Does the axiom refer to internal states of mind, verbal and written expressions of preference, or binding commitments to particular choices? The modal flatness[3] of mathematics seems to lie at the heart of this interpretational quandary. Phrases such as 'there exists a mapping . . .' can be useful to

---

[2] A BBC2 game show in which one team gives three possible meanings for a word and the other has to guess which is right. Teams swap roles for every word until a time limit is reached. Surprisingly gripping stuff.

[3] The term is due to J. R. Lucas.

mathematicians without too much ontological delving, but such questions cannot so lightly be skimmed in mathematical applications. At the moment of interpretation, decisions must be made.

We noted that the two accounts of completeness presented above are often treated as being formally equivalent, although that due to Savage is preferred for its economy (one primitive relation, R, instead of two, P and I). However, before we evaluate particular interpretations, it is useful to note a sense in which the versions of the axiom are not equivalent. To interpret this or any other axiom, we might adopt the following device: first assume that the axiom has some content, and then outline the conditions which, if they held, would count as a violation. If we use the two definitions of completeness above to derive sets of falsification conditions, we obtain

$$V_{vNM} = \{Pab \wedge Pba, \; Pab \wedge Iab, \; Pba \wedge Iab, \; \neg Pab \wedge \neg P\text{-}ba \wedge \neg Iab\}, \tag{3}$$

and

$$V_S = \{\neg Rab \wedge \neg Rba\}, \tag{4}$$

where V_{vNM} and V_S are the sets of conditions that count as violations of completeness under the von Neumann–Morgenstern and Savage definitions, respectively. In Savage's system, three possible primitive statements concerning binary relations ($Rab \wedge \neg Rba$, $Rba \wedge \neg Rab$, and $Rab \wedge Rba$) are consistent with completeness, which means that there is only one way ($\neg Rba \wedge \neg Rab$) in which the axiom could be violated. The von Neumann–Morgenstern notation and definition give a richer set of possible conditions for axiom violation. Though less economical, this version has more falsifiable content. The point seems to have been overlooked, though its implications for the way in which axiomatic models are evaluated are not unimportant. Even if we operationalize behaviour in such a way that for any experiment the contents of (3) and (4) were identical, the von Neumann–Morgenstern version provides more help in trying to understand why violations of completeness might occur. And because the ability to explain is an important element of normative judgements, it is their version that will be used to underpin our discussion of completeness.

Preference Formation and Counter-Preferential Behaviour

As the von Neumann–Morgenstern primitive relation indicates that conflicting preferences (e.g. *Pab* ∧ *Pba*) may provide a source of completeness violation, it is natural to ask what conflicting preferences mean and what grounds we have to suppose that they exist. Economics is better at handling such issues than it was even a decade ago, though substantial literatures elsewhere have a longer tradition.

A good proportion of the philosophical literature on counter-preferential behaviour is devoted to determining whether the notion is coherent; if it were not, incomplete preferences would be contradictory. Unfortunately, the debate is not conclusive, and draws on the kinds of arguments that are only just beginning to make their way into economics through game theory as well as from philosophical decision theory. For both these reasons, interested readers are referred to discussions elsewhere (in Elster 1983; Levi 1984; and Morton 1991), and we shall move on to an alternative approach that derives from the social psychology literature on attitude-discrepant behaviour.

There are, in fact, many theories that seek to explain how attitudes (preferences) are formed and to identify conditions under which they change. Here we shall consider only three. The first, known as functionalism, due to Katz (1960), has as its basis the idea that attitudes do not just serve to drive behaviour. For instance, attitudes help us come to conclusions in the absence of perfect information from which these might be derived. They serve a number of social functions by allowing us to express central values, define ourselves, and relate to particular groups. They may even be used to defend egos against facts about which we feel uncomfortable. By this account, attitudes are a resource in a cognitive economy which is used to do things other than just produce behaviour, so there is no reason to find the one-to-one mapping between attitudes and behaviour that completeness (so construed) requires.

A second theory, known as cognitive dissonance, due to Festinger (1957), emphasizes an internal need for consistency that pulls attitudes and preferences together. The theory would seem to suggest that conflicts between preferences and behaviours will be eradicated over time. However, the classical experiments in this area were based on payments to subjects who were asked to perform unexciting tasks and were then paid either high or low amounts depending on the experimental variation. Those who were given the higher payment ranked the tasks as being duller, a finding the theory had predicted. Subjects paid only a small amount rationalized their participation in the experiment by raising their rating of the interest of the tasks set. Cognitive dissonance suggests not that behaviour and attitudes will differ, but rather that they will converge (with completeness), but for reasons that are hard to describe as rational.[4]

A third theory of attitude formation and change, originating in part from the methodological debate that surrounded work on cognitive dissonance, was proposed by Bem (1972). Bem's thesis was simple. Instead of viewing preferences as motives for behaviour, he suggested that attitudes might be the inferential *products* of action. What we know about our likes and dislikes, we infer from observing and recalling our behaviour as much as by introspection. Such a theory is certainly friendly to the *animus* of neo-classical economic behaviouralism, and in reversing the presumed causality between attitudes and behaviour, Bem was, in effect, revisiting the idea of logical behaviouralism, discussed extensively by the philosopher Gilbert Ryle (1949). Possibly carried away by his own prose style, Ryle might have pushed the argument too far, but the basic point—that attitudinal talk is partly a metaphorical cover for what we do—must surely be right. There is some question whether this view leaves much space or need for preferences themselves—or for any other metric of welfare—but, in so far as it does, it casts violations of completeness as the consequences of inadequate information and thereby renders them impotent as challenges to rationality.

---

[4] An interesting application of the theory to the use of protective clothing is reported by Akerlof and Dickens (1982).

Extensionality

While classical decision theory excludes the possibility that the particularities of equally informative descriptions affect behaviour, it is not clear where this assumption resides. This issue does not arise in the special case when the domain of utility is money, but it does when a wider class of consequences is allowed. It is possible to say that consequences cannot include the non-informational aspects of state descriptions and to regard the assumption as a domain restriction—albeit one that is usually implicit. Alternatively, since description-sensitive choice (driven by intensional preferences) provides the possibility of observing preference reversals—*Pab* under one description and *Pba* under another—it might well be associated with the completeness axiom. Either way, the assumption is one that SEU makes, though not one that is always met, as an experiment discussed by Arrow (1982) illustrates. The experiment concerns clinicians' recommendations for surgery based on information about the survival probabilities. When told that the chance of survival was 0.9, some 80 per cent of surgeons who were asked recommended surgery. However, of those told that the same course of action was associated with a 0.1 probability of death, only 50 per cent recommended surgery. The informational content of each phrasing is identical, so it seems as if some surgeons must have had preferences that were intentional.

This kind of finding is the very stuff of opinion pollsters and market researchers. The mere fact of such behaviour does not in itself prove rationality, although by throwing light on the motivation we may begin to address the normative question. Three explanations seem particularly relevant. First, surgeons could have preferences for particular numbers, and a proposition with a '9' in it may be more desirable than one with a '1' in it—such preferences can be observed in societies ranging from primitive tribes to stock-market traders.[5] Second, it is possible that the presence of particular words such as 'survival' and 'death' make the outcomes to which they refer more salient. Third, it could be

[5] J. Cohen (1960) reports a young girl whose lucky number was 2 because that was the number of children she wanted.

that the observed description sensitivity reflects a socio-linguistic norm that surgeons use when making recommendations to patients or colleagues. If a doctor wants to recommend surgery, the convention might be to make explicit the positive outcome from that course of action; if the counsel is to avoid the operation, then the negative outcome of the operation is the one expressed explicitly.

Taken at face value, the first two explanations point to determinants of choice that we might find it difficult to accept on normative grounds. Even if surgeons do have intrinsic preferences for particular numbers, it is difficult to see that patients would want this to have an impact on the choice of treatment for them. Much the same can be said for the salience thesis. Although there exist non-cooperative games in which focal outcomes can be used to select strategies optimal for both parties, the situation above is one in which salience serves no such purpose.[6] *Au contraire*, salience here has the feel of introducing a bias that has no bearing on what the rational agent must do.

The third hypothesis, namely that the results reflect a convention in everyday use by surgeons, though it begs the question as to why such conventions are used, does not of itself establish irrationality. If a clinician wanted to recommend surgery, making the positive (survival) outcome focal by uttering it explicitly is a reasonable semi-conscious way of signalling and reinforcing this. (Indeed, it could be regarded as indicating the ecological validity of the experimental design and the naturalness of the responses obtained.) I do not want to argue here whether the findings Arrow discusses are necessarily rational, though I do think we can say that the social-convention argument leaves the normative case against the intensional less clear-cut than we might suppose.

Completeness as Precision

Completeness with respect to any pair of options can be seen as excluding any sense of vagueness about one's own preferences.

---

[6] Bacharach (1991) provides a theory that shows why (and when) picking the salient action is rational.

For example, you might definitely prefer three small bottles of mineral water to a pint of stout and yet, because you were thirsty, prefer the pint if the alternative was only one bottle of water. If the choice were between the pint and two bottles, you might not be sure how you would choose—you might dither when choosing. Here it seems more appropriate to say that the violation of completeness is due to imprecision rather than to incommensurability.

In one light, at least, precision is basic to rational choice theory because it is the metric of preference by which risk is assessed. This is certainly true in social situations where accountability is important. Politicians and civil servants alike know, for instance, that the more precise are the criteria for ranking outcomes before the event, the more likely they are to seem, *ex post*, to have failed. The converse may also be true. For instance, individuals may seek to make their preferences precise as a form of target-setting designed to enhance their own performances. The literature shows the motivational efficacy of goals that are neither too easy nor too ambitious. In short, the reasons both for and against the precision interpretation suggest that, although completeness may have a normative justification, the appeal is contingent.

In Savage's scheme, the only time in which completeness is violated is when we observe $\neg\, Rab$ and $\neg\, Rba$, a condition commonly associated with problems of incommensurability. If the primitive is preference, there is no difficulty in allowing that we cannot sometimes make up our minds. The position advanced here is a mild revision of the position I took earlier, but it still indicates that commensuration is not the only issue on which completeness turns. In some interpretations, violations (intensionality) look like the results of cognitive bias; in others (precision) there seems to be a case for violations; and in a third scenario (where cognitive dissonance features) it seems that cognitive pressures will tend to ensure that attitudes and behaviour cohere, though the sensation of those pressures may not, in itself, be pleasant. It is beyond the scope of this chapter to discuss the wider implications of this, but it is fitting to end by noting how integral to decision-making incompleteness is. This is well understood by decision analysts, who use SEU to support decision-making, and who spend much of their time not merely eliciting preferences, but helping decision-makers to construct objectives. If completeness held in practice, many of the issues that make decisions problematic, the estimation of trade-offs being one, would disappear. As the late George Shackle (1961: 4) put it, decision-making would become 'a mechanical and automatic selection of that act whose outcome [the individual] most desires'.

CONTINUITY

Many of the theories that generalize SEU by relaxing independence retain the assumption that utility is continuous with respect to the probabilities. Suppose there are three outcomes, where h and l are the most and least preferred and m is an intermediate outcome. Continuity (according to von Neumann and Morgenstern 1944: 26) requires that m is preferred to a probability mixture of the extremal outcomes for some sufficiently small, strictly positive, probability of h. Likewise, the probability of l can be made so small that a lottery offering the h and l is preferred to a sure payment of m.

Continuity is an assumption that some would argue has only technical content and should not, therefore, play any role in the normative evaluation of SEU. In the absence of any principled distinction between technical and other kinds of content, it is difficult not to be sceptical. If we were to accept that continuity is merely a technical feature of differentiable functions, then why do we not similarly accept that transitivity is a technical feature of uni-valued functions?

Finally, we look at two attempted defences of continuity—readers may judge for themselves whether these suggest that completeness is merely formal. The first line of defence, which emphasizes just how small non-zero probabilities can be, is a response to a *reductio ad absurdum* proposed by Alchian (1953). Suppose h, m, and l are 'win £5', 'do nothing', and 'be shot in the head'. Continuity requires that we must, for some non-zero probability of being shot in the head, take a lottery in which the other outcome is 'win £5'. Alchian questions whether such preferences are really mandated by rationality alone and his doubts are easy to share. On the other hand, Green (1971) proposes that we consider a modified lottery in which the worst outcome is to 'be shot at from 1,000 miles'. In so far as Green's modification reduces the probability of the lethal outcome, it is indeed likely that more people would take the risk. However, it is still far from clear that you *must*, on pain of being judged irrational, take such a risk. Another suggestion, which can be found in Ng (1975), is that probabilities such as $10^{-10^{-10}}$ are just not worth bothering about. If completeness has substantive implications, they are innocuous. As an answer to the question, How should decision-makers go about the business of practical deliberation? there is little reason to disagree with Ng's proposal. However, this satisficing-based advice depends on an acceptance of the fact that cognitive powers are finite, and it is difficult, therefore, to see how it might ground a theory of utility *maximization* and address questions concerning the appropriate behaviour of idealized decision-makers.

The second attempted defence of completeness likens the example to risks taken as part of everyday life. People cross roads to buy newspapers, so they often do bear small risks of fatality for moderate gains. As long as the gains and probabilities of losses are at least as favourable as those encountered in reality, why should being shot in the head be any different? Even if the point has some merit, it is insufficient to provide the universal support that an axiom purports to have. Those decision-makers who accept the equivalence of Alchian's decision problem to ones that we encounter every day are precisely those who want to strip down the problem descriptions to the level at which probabilities and 'benefits' appear to be similar. However, any rational agent who wants to reject continuity in Alchian's problem will reject description-coarsening because it removes precisely the differences that provide the warrant for different behaviour.

In concluding this discussion of continuity and completeness, it is worth noting that they seem to reinforce certain lessons learned from considerations of transitivity and independence. First, the appeal of an axiom seems to depend on how the relation whose structure it constrains is interpreted. The convention of leaving relations primitive, therefore, leaves the normative question under-identified. Second, how choice options are described seems also to be linked to the normative status of the axioms that control relations. In the following chapter, I shall reinforce the importance of both issues by giving them a more formal treatment and considering some methodological issues that classical decision theory raises.

7

Interpreting Axiomatic
Decision Theory

INTRODUCTION

The axiomatic method on which modern utility theory stands
has been used by economists and others to study social phenomena, ranging from optimal choice under risk, game theory
(Zermelo 1912), social choice (Arrow 1951), moral philosophy
(Harsanyi 1955), and general equilibrium to poverty (Sen 1976).
Although this approach has become widespread since the 1960s,
it has a history that stretches back over at least fifty years and
might plausibly be traced back to Ricardo's preference for
deductive modelling—the so-called Ricardian vice.

Typically, modern axiomatic theories, particularly those in
utility theory, employ a primitive, ordinal relation and seek to
derive representation and uniqueness theorems from statements
about properties of the relation. Researchers in a number of
areas (some already mentioned in previous chapters[1]) have suggested how the primitive entities of certain choice theories should
be construed, though it is still standard practice to leave this
question open. We say of preference, for instance, that $(a)(b)(Rab$
$\vee Rba)$, to state that the relation R is complete without saying
very much about the substantive nature of the relation R
(apart from the rather cryptic remark that it denotes '*the* weak
preference relation'). Perhaps, for consistency's sake, we say

---

[1] Not covered here are the remarks about the interpretation of primitives in
social-choice theory. Or particular interest are papers by Kelsey (1987) and
Gaertner *et al.* (1988), which identify problems arising out of formal theories of
social choice that do not sufficiently specify the entity domain over which rights
can be exercised.

little also about the rules for individuating entities over which R operates.

The practice of proposing axioms that are at most only partially (or un-) interpreted is convenient, but is it innocuous? The question arises *par excellence* in experimental tests of axiomatic theories of utility, where the consequences derived from the theory must be operationalized, but it also arises as a matter of general concern. Practice suggests that most researchers believe the answer to the question with which this paragraph began should be given in the affirmative.

In this chapter I shall discuss more formally, and in a more methodological vein, some of the issues that arise from attempts to 'individuate primitives' (as Broome 1991 puts it). Most of what I shall say turns on the distinction between the way primitives are used in pure mathematics and the way it is hoped they might be applied to decision-making. I shall add to those arguments that have been advanced against this consensus by drawing attention to some of the problems that befall the usual claims of normative content made for axiomatic theories.

MATHEMATICAL AXIOMS AND NORMATIVE APPEAL

In what way do mathematical axioms help establish the normative appeal of particular models? The literature is far from clear about this, but I shall propose two views which seem to underlie the discussions that do exist. The first view, call it the *logical view of normative appeal* (LVNA), is that normative appeal is transmitted through logical relations. Because a solution (often a representation and uniqueness theorem) is *logically* derived from a set of axiom propositions, it assumes the normative appeal of those axioms. After all, what better links could we expect than those on offer from first-order logical entailment?

Appealing as LVNA may be, it could never be complete. The first cause for concern arises when we recall that logical relations can be problematic transmitters of properties in which we are interested. Hempel's paradox, for one, shows that what counts as verification of a universal proposition does not necessarily count as verification of logically equivalent propositions—white swans do not count as evidence that ravens are black.

The problem is this. Superficially, it seems reasonable to all that any evidence for, or against, a proposition should count for, or against, all logically equivalent propositions. But suppose we have evidence which we decide is relevant to the proposition, 'all ravens are black'. This is equivalent to another proposition, namely, 'all non-black things are non-ravens'. Now suppose we find a white swan and allow that that is relevant to the proposition that all non-black things are non-ravens. The equivalence rule, which seemed sensible, forces us to take the observation of a white swan to be counted as relevant to the original proposition. Most people think this is absurd.

The point was made in the context of proposals for the conduct of certain types of positive sciences, but it demonstrates the more general point that logically equivalent propositions are not co-extensive. It would be wrong to draw on some covering law about the property preserving nature of logical relations, for no such law exists. Perhaps there is a special case which can be made for normative appeal, though I know of none. Such an argument would show that the implication connective between consequents and antecedents is sufficient to link their normative status. Any 'if and only if' theorems would show both that solution concepts (like 'maximize expected utility') derive appeal from axiomatic propositions (like connectedness, transitivity, independence) and vice versa. Some axiom set would be appealing because it was a logical consequence of some solution, a claim that fails to model the asymmetry of standard, normative appeal claims *from* axioms *to* solutions. We do not, for instance, advocate transitivity on the grounds that it can be logically deduced from expected utility maximization.[2]

If there are problems with LVNA, then it is natural to ask if these derive from the fact that the apprehension of normative appeal, and therefore the *raison d'être* for employing axioms, is

[2] Indeed, it seems to be an important feature of impossibility theorems that they show how assumptions that are individually appealing have no such appeal as a collective. I am grateful to David Kelsey for this point.

essentially cognitive—a view we might call the *psychological view of normative appeal* (PVNA). Under this view (one that Plato appears to advocate in the context of both general and mathematical (geometrical) argument), we should employ axioms because they describe propositions that we could easily accept. Indeed, it is easy to accept that single axioms considered in turn may be easier to evaluate than the implications of a particular solution.

This view seems a closer fit for the account that we expect axioms to play in the partially interpreted theories of modern economics. But if normative appeal really does reside in particular descriptions, if it is an intensional concept, then we are forced to ask how our knowledge of logical relations is relevant. If the normative appeal resides in the particular description of a proposition, then different descriptions—even of the same propositions—will have different normative appeals. Even if we assign to a proposition the appeal of a modal (average) description, this would hardly provide support for the use of axioms under the particular limited set of descriptions most theories generally employ.[3] While PVNA gives weight to cognitive factors that are omitted by LVNA, it provides no grounds for the extensionality on which the use of axioms is based.

There is, it seems, a question about whether we can give an account of the normative appeal of certain solutions to decision-theoretic problems by grounding them axiomatically. The conclusion may seem rather sceptical, but results in the new decision theory often do seem to require that we re-evaluate our notion of optimal decision-making at a fundamental level. For instance, in his review of the normative arguments for non-expected utility theories, Machina (1989) has pointed out that we can no longer accept a priori that sunk costs are irrelevant in decision-making. Developments in the mathematics of utility theory, he claims, show that whether or not fixed costs bear upon the decision-maker's short-term calculations depends on the separability of the utility function.

---

[3] In practice, theories are usually presented with two descriptions of axioms: one mathematical, and one more or less in the natural language.

AXIOMS, MEANING, AND CONTENT

Suppose that some complete, non-self-defeating account of normative appeal could be given: what then could we say about
axiomatic theories? Could we say that transitivity was a constitutional element of rationality, so theories of rational choice must
constrain behaviour to be rational? This argument has been
addressed already in Chapter 4, but I want to refute the claim,
not on the grounds that the arguments for it are weak, but
rather that they are empty. In particular, I shall argue that
transitivity is a matter of language, not of behaviour—a view
summarized in the following theorem:

> TRANSLATION THEOREM   All intransitive behaviours can be
> redescribed in such a way that transitivity is not violated,
> and all transitive behaviours can be redescribed in such a
> way that transitivity is violated.

Remark   Whether or not transitivity is violated depends on the
description of the behaviour used, not on the behaviour itself.

Proof   First, to show that all intransitive behaviours can be
given a description in which transitivity is not violated, we
assume that an apparently intransitive set of pairwise choices
over a, b, and c can be given an intransitive description as in
(1)–(3):

$$Cab \tag{1}$$
$$Cbc \tag{2}$$
$$Cca \tag{3}$$

where a, b, and c are some objects over which an agent can
choose and C is the two-place predicate 'is chosen over'. Now
suppose we refine the primitives so that they indicate the choice
sets of which they are part and then describe the *same* behaviour.
The following set of statements is obtained:

$$C(a \text{ out of } a \text{ and } b)(b \text{ out of } a \text{ and } b) \tag{1'}$$
$$C(b \text{ out of } b \text{ and } c)(c \text{ out of } b \text{ and } c) \tag{2'}$$
$$C(c \text{ out of } a \text{ and } c)(a \text{ out of } a \text{ and } c) \tag{3'}$$

Whether this new description system is more refined in Sen's sense of a refined description (Sen 1987) or merely contains more redundancy than the original system depends on what is understood by propositions such as *Cab*, but the point to note here is that (1')–(3') describe the intransitive pairwise choices with which we started.

Now suppose a new linguistic convention is produced in which the terms above are translated as follows:

Initial descriptions	*New linguistic convention*
(*a* out of *a* and *b*)	*l*
(*b* out of *a* and *b*)	*m*
(*b* out of *b* and *c*)	*n*
(*c* out of *b* and *c*)	*p*
(*a* out of *a* and *c*)	*q*
(*c* out of *a* and *c*)	*r*

The two conventions are linked by a one-to-one mapping so there need be no difference in the information conveyed in (1')–(3') or by the following statements:

Clm	(1″)
Cnp	(2″)
Crq	(3″)

Using the new linguistic convention, we obtain three statements about six primitives and the intransitivity is dissolved.

A similar translation can be made from transitive to intransitive choices. To see this, we can follow a similar set of steps. First, assume a transitive description:

Cab	(4)
Cbc	(5)
Cac	(6)

Refinement yields the following set of propositions:

C(a out of *a* and *b*)(*b* out of *a* and *b*)	(4')
C(b out of *b* and *c*)(*c* out of *b* and *c*)	(5')
C(a out of *a* and *c*)(*c* out of *a* and *c*)	(6')

Now construct a linguistic convention in which new primitives are assigned as follows:

Initial descriptions	New linguistic convention
(*a* out of *a* and *b*), (*c* out of *a* and *c*)	*m*
(*b* out of *a* and *b*), (*b* out of *b* and *c*)	*n*
(*c* out of *b* and *c*), (*a* out of *a* and *c*)	*p*

In the new language, for all instances of *a* out of *a* and *b*, and *c* out of *a* and *c*, we write *m*; and so on. We rewrite (4)–(6) using the new convention thus:

$$Cmn \qquad (4'')$$
$$Cnp \qquad (5'')$$
$$Cpm \qquad (6'')$$

Under this system of description, the behaviour is intransitive. Q.E.D.

Remarks Without prior agreement on the linguistic conventions that will be used to say what counts as a particular choice primitive, we can choose, *ex post facto*, some convention (richness of language permitting) in such a way that an observation (set) can be counted, either as a violation of transitivity or any other axiom that we are testing,[4] or not, depending on choice. Axioms like transitivity (though not completeness) require more than one instance of a particular stimulus, and without some criteria as to what these are, our decision theory will have no content. Although a (relatively) formal proof has not been used before, similar arguments can be found in work on philosophical psychology (Davidson 1980; Tversky 1975) and in economics (Broome 1991). In addition, there are parallels with the problem in Frege's naive set theory (Frege 1893) which Bertrand Russell identified when he showed that, without constraining what is to count as a set, we can obtain a contradiction—the finding now

---

[4] Strictly speaking, this proof applies only to transitivity. However, the discussions in the literature referred to in Ch. 5 seem to suggest that the issue applies also to the independence axiom. I have an intuition that the point does not apply to completeness, but I cannot yet prove this.

referred to as Russell's paradox.[5] In decision theory, without
some constraint on what counts as a choice primitive, the theory
has no content.[6]

We know that, using an axiom devised by Zermelo (1908),
basic contradictions of Fregean set theory can be avoided. We
may therefore ask whether a similar move is not also open to
the decision theorist. In fact, two such moves are used in the
literature, but both are rather costly (for different reasons) when
measured in terms of their effects on the objectives that decision
theorists set out to achieve.

The two proposals for entity individuation I have in mind
may be called 'consequentialist' and 'spatio-temporal'. The consequentialist procedure is most forcefully argued for by Savage
(1954: 9 *et seq.*) and more recently by Broome (1991). If the
agent derives utility over some attribute—if it is of concern to
the individual—then, whatever it is, it should be modelled as
part of the utility-yielding primitive. So if, for instance, *a* out of
a and *b* gives more or less utility than, say, *a* out of *a* and *c* for
the agent, they are recorded as using different primitives. Similarly, if the common consequences that the independence axiom
rules out of consideration matter to the decision-maker, then the
consequentialist procedure for constructing primitives rules them
back in. Such liberalism would finesse the possibility of testing,
and the remarks that Savage makes about *violations* of his
axioms indicate that he never acknowledged the full (nihilistic)
consequences of his pre-axiomatic proposals for primitive construction. (On this theme see also Bausor, 1985.) To sum up,
consequentialism seems to satisfy Hume's postulate, but at the
expense of testable, behavioural content.

[5] If there were a universal set, it would contain all other sets including the set
of sets that are not members of themselves. If a set is a member of this set then it
is not a member of itself, so it cannot be a set. If a set is not a member of itself
then it should be in the set of such sets, though this cannot be the case, for all
members are sets and therefore members of themselves.

[6] A similar point is made by Sen (1979) in his attack on the Pareto principle
which has become a cornerstone of welfare economics. The thrust of Sen's
argument is that, while the principle is acceptable as a dominance argument, it
fails as a universal social choice rule because its application to particular choice
situations permits choices that (for reasons such as concern for human liberties)
we might wish to disallow.

So what of the alternative, the spatio-temporal approach? Following Debreu, this bases choice on option descriptions characterized by the ordered triple $\langle p, l, t \rangle$ where p denotes physical properties, l the (geographical) location, and t the time at which the commodity becomes available. Unless we believe that rational agents should be concerned only about the physical properties, locations, and 'dates' of commodities (and not even utilitarianism enforces such relentless materialism), then the spatio-temporal primitive will quickly fall foul of Hume's postulate. With very long time periods (and the ability to replicate p and l characteristics), it seems reasonable to suppose that we could find repetitions of particular primitive objects and that we could, therefore, test any axiom we chose. However, if, over the course of a year, a person is given three binary choices and seems to violate transitivity (according, that is, to linguistic conventions which satisfy the Debreu rules for primitive construction), then in what sense would we want to say that the person has been irrational? People and environments change. Clearly a year is too long. How about a month? Would a violation of transitivity within a month count as irrational? We can continue in this vein, but it is difficult to see where exactly the axiom will bite. Will it be 3.2 minutes or 0.0107 micro-seconds, or . . .?

Quite apart from failing the 'hands-off' test, it is not clear that the spatio-temporal approach can deliver axiom interpretations that have any content. Such interpretational issues can have ramifications that are widespread. For example, it is possible to criticize the systematic error approach to axiom validation. Any set of errors can be given many non-uniform partitions, so any set of errors can, under some interpretation, be regarded as 'systematic'. Formalists cut themselves off from ways of considering only systematic errors that have an interesting interpretation.

IMPLICATIONS

In this chapter, I have questioned the claim that mathematical–economic theories like SEU have normative content, on the grounds that the mechanism by which normative appeal is transmitted through logical relations is problematic and that axioms *per se* do not have any content. An implication of our argument is that, for axioms to have content, they must be accompanied by rules for constructing the primitives over which they (the axioms) are defined.

In order to see what would constitute a set of sufficient conditions, it is helpful to consider a simple example. Suppose that L denotes the relation 'is strictly greater than' and that there exists a domain, $\{7, 5, 1, 0\}$. Then $\langle L, \{7, 5, 1, 0\}\rangle \in \mathscr{C} \cap \mathscr{T}$, where $\mathscr{C}$ is the set of numerical structures with a connected relation and $\mathscr{T}$ is the set of numerical structures with a transitive one. If we altered the relation to L^*, defined as 'is greater than by more than 1.5', i.e. $(L^*xy \Rightarrow Lxy)$, the numerical structure has a different axiomatic profile; in particular, it is no longer an element of $\mathscr{C}$. If we altered the domain to $\{7, 5, 1\}$, then the structure will again be a member of $\mathscr{C}$. The moral is that whether axioms hold depends crucially on (i) the domain of the relation and (ii) the meaning of the relationship itself. The unit of content is not the axiom but the relation–domain pair, and debates about the normative appeal of an axiom without reference to the relation–domain pair should puzzle us. The lesson is an old one, but if it holds for the relatively uninterpreted world of pure mathematics, then it would seem to hold, *a fortiori*, for mathematical–economic theories.

The problem, I believe, is not that our procedures for primitive construction are defective, for they are plausible responses to different problems in decision theory. If, for example, the theory is to be testable, and if this entails that the theory must make non-empty behavioural predictions, then rules for individuating primitive objects and identifying the admissible domain of those objects must be produced. In addition, primitive relations must be given a substantive meaning. When fully dressed in such a way as to give content, a classical decision theory is an ordered quadruple of the form $\langle \mathscr{R}, \mathscr{A}, \mathscr{T}, \mathscr{D}\rangle$, where $\mathscr{R}$ is the set of rules for interpreting the primitive relations, $\mathscr{A}$ is the axiom set, $\mathscr{T}$ is the set of individuation rules, and $\mathscr{D}$ is the domain over which the primitive relations are defined. This decomposition of mathematical decision theories maintains the Debrovian division between the interpretation and technical parts of the theory, but I think it would be a mistake to regard some elements as being more important than others.

To put this another way, we might say that for transitivity to have content it requires agreement on linguistic conventions, and for the axiom to have normative content, the linguistic conventions would have to be the conventions of a rational language. I doubt that such a notion is philosophically coherent, and Hirsch (1988) seems to support this case. A fascinating reformulation of game theory in terms of metrics over possible worlds–space (Shin 1989) brings to mind a mathematical objection also. If there were a non-null set of rational languages, it is most unlikely that it would be a singleton.

An alternative use of SEU, which may seem paradoxical, given that theories are often regarded as an escape from facts, is its use as a descriptive tool. Here decision theory is being used to infer, from a person's choices, his or her desires (in revealed preference theory) and beliefs (in the more general theories of decision-making under risk). Now it should be clear that to solve this problem primitives are required which reflect everything that is of consequence to the decision-maker. But how do we know what that is, especially if we disagree with Becker's claim that all decision-makers have the same preferences? Using decision theory to solve the descriptive problem, it is possible, as the Translation Theorem shows, always to reinterpret the primitives so that axioms are not violated. In this process, where decision theory is being used purely as a measure theory, rather than as some aid to decision-making, the sets $\mathcal{T}$, $\mathcal{R}$, and $\mathcal{D}$ are not posited prior to the experiment, but rather emerge *ex post*, and they will not typically be unique, at least for finite experiments. Whether, for instance, preference is transitive or not is answerable only with respect to specified languages.

Both projects seem to be reasonable ones to follow, but it is wise to be clear about the extent to which the primitive materials required by the projects differ and the nature of those differences. The fact that the 'mathematics' is the same in either case masks the difference of the strategies appropriate to the construction of the primitives. Solving *simultaneously* for the twin objectives of being 'hands-off' and providing content is like trying to square the circle.

8

Uncertainty and a New Axiomatic Theory of Rational Choosing

WHAT'S WRONG WITH SUBJECTIVE PROBABILITY

If uncertainty has been a major source of debate for economists this century, then their serious exchanges have been conducted almost wholly in the language of probability. There are, however, a number of proposals for describing and modelling beliefs in situations where probabilities are known with different levels of certainty, or not known at all. These alternatives include the measure of *qualitat* in Meinong (1890), *sicherheit* in Nietzsche (1892), *weight* in Keynes (1921), *potential surprise* in Shackle (1952, 1961) and Levi (1966, 1984), *acceptability* in Barnard (1961), *ambiguity* in Ellsberg (1961), and *epistemic reliability* in Gardenförs and Sahlin (1982). Even computer scientists, for example Shortliffe (1976), have devised expert systems which use non-probabilistic *certainty factors*.

Occasionally questions surface as to whether these alternatives to probability are suitable for normative purposes. Each time, these doubts are either discouraged by Bayesian counters or ignored on the grounds that the theory we already have seems sufficiently powerful. To understand why the neo-classicist's answer to this question is negative, it is helpful to follow the history of the risk–uncertainty distinction in economics, which is not widely available to decision theorists from other disciplines.

Finding a model of the probability axioms in the absence of relative frequencies has long been an unresolved issue in the foundations of statistics. Proposed solutions include the Laplacean use of equi-probability in the face of insufficient reason to think otherwise, and the idea that there is an *ex ante* propensity for (even unique) events to occur (von Mises 1957). Knight is known for crystallizing this problem within the context of economic analysis by means of a distinction between risk, where objective probabilities exist, and uncertainty, where such probabilities are not available. As noted elsewhere (Anand 1986), it is clear that Knight saw the distinction as being of practical significance:

The businessman himself not merely forms the best estimate he can of the outcome of his actions, but he is likely also to estimate the probability that his estimate is correct. The 'degree' of certainty or of confidence felt in the conclusion after it is reached cannot be ignored, for it is one of the greatest significance. (Knight 1921: 226)

Knight's distinction is a precursor to two developments. One provides axiomatic theories of choice based on knowing only the consequences of states without making any assumptions about the different probabilities that those states will occur. No substantial critiques of this approach exist, though it has not found favour with applied economists who feel that the assumption of complete ignorance is too extreme. The alternative approach sees the subjective probability framework, due to Ramsay and de Finetti and popularized by Savage, as delivering subjective probabilities in the absence of objective ones. After all, what other probabilities could we expect in the face of uncertainty?[1] An extension of this view is[2] well entrenched in the literature, as can be seen in graduate texts (e.g. Deaton and Muellbauer 1980), and is succinctly summarized by Borch:

the convention [Knight's] does not serve any useful purpose either theoretical or practical ... because, since Savage, all probabilities are subjective. (Borch 1968)

---

[1] Recent work by LeRoy and Singell (1987) contrasts with the received view of what Knight meant and indicates that he was, in fact, a supporter of subjective probabilities.
[2] Views about the normative status of SEU are changing so quickly that we are on the verge of having to replace 'is' with 'was'. Kreps's micro-text certainly shows a deep awareness of the normative objections to SEU.

Either way, the claim that the distinction is otiose predominates, although the critical Ellsberg (1961) experiment undercuts the basis on which the distinction might be dissolved. Ellsberg's own results are the products of informal experimentation and might reasonably be questioned—by the standards of current practice they certainly would be. However, his major achievement was in designing the experiment, and showing how a person sensitive to uncertainty might have beliefs that could not be described as subjective probabilities. Given the importance attached to the ability of subjective probability to fill the gap that relative frequency cannot, in the absence of information, it may be surprising that more attention was not given until recently to Ellsberg's work. However, we should note that he makes no comment on the normative status of the behavioural violations of SEU that he finds, though this is a key issue for statisticians—even more so than for economists.[3] All the major tenets of classical decision theory are currently up for reappraisal, and Ellsberg's ideas have been central to our revisions of concepts of belief; a normative defence of his position can be found in Gardenfors and Sahlin (1982), Hay (1983), and Anand (1986), and measure-theoretic results based on his ideas are provided in Fishburn (1983b). The only substantial departure from Ellsberg's view that hindsight allows is that the conceptual basis for behaviour under uncertainty seems to owe rather more to a brief but intriguing chapter in Keynes's (1921) *Treatise on Probability*, on weight of evidence, than it does to Knight's risk–uncertainty distinction.

In his *General Theory*, Keynes (1936) notes (literally) that what he meant by uncertainty in the macro-economy is to be associated with weight. We may take it, then, that Keynes came to believe that weight and uncertainty were crucial to economics, though it is even clearer that when he wrote the *Treatise on Probability* he was far from sure what use the construct of weight might have. J. L. Cohen (1985) makes proposals that build on Keynes's ideas (which are themselves elaborations of

---

[3] Normative concerns are embedded in the constitution of statistics, whereas they are only interests in economics, albeit powerful ones.

remarks made by the German philosophers Meinong (1890) and Nietsche (1892)), while Runde (1990) draws our attention to the different meanings that Keynes ascribed to weight. At one moment he refers to the 'amount of relevant evidence' in such a way that suggests a measure of the quality and quantity (e.g. sample size) on which a probability estimate might be based. Elsewhere, however, he writes of the balance of absolute amounts of relevant evidence and relevant ignorance on which a probability is based. A third definition associates weight with the 'degree of completeness' of information on which a probability is based, though Runde takes this to be equivalent to the second definition. The view I advocate (which Ellsberg 'proves') is that an epistemic measure of information quantity and quality is what decision theory most needs in addition to the probability calculus, and I therefore use the term 'weight' in that particular sense.

To capture this second dimension of belief, I propose using a variable, ω, which measures the weight of evidence. It is possible, then, to think of agents either as consciously maximizing with respect to some set of objectives or as behaving as if they were maximizing the value of a particular function, f, which is defined over probabilities of consequences in all possible i states, $\{p(c_i)\}_i$, and weights of probability functions, $\omega(\{p(c_i)\}_i)$, or $\omega(p)$ for short. The desirability of an act is now given by the sum of $f((p_i,c_i), \omega(p))$ over all i states, rather than simply $f(p_i, c_i)$. It is harder to be exact about the substance of ω (what values it can take and how we know what value it does take for any particular distribution), but recall that similar problems arise with the probability measure. Some physicists, for instance, have found uses for negative probabilities (Feynman 1987); there may be little difference between this and the use of imaginary numbers in engineering or economics, but many would reject negative probabilities.

In general, the statistic ω may not be easy to calculate. In situations where relative frequencies are being used to provide estimates of probability, it would be natural to take n, the number of observations on which the probability estimate is based, as an estimate of the quantity of information. In any case, it should be evident that the two-dimensional view allows one to distinguish between identical relative frequencies estimated from different sample sizes, which, strictly speaking, the one-dimensional theory of probability theory does not. (Hypothesis-testing takes account of information quantity, though only via the satisfaction of arbitrary 'significance' levels.) In practice, there may be a variety of instruments that one could use to evaluate a person's subjective estimate of the quantity and quality of information. While operationalization is important, I shall not consider the technical or methodological issues involved in inferring values of ω or measuring attitudes towards it. Rather, I consider the normative issue, that is whether the notion of rational behaviour is sufficient to establish the need for a two-dimensional concept of probability. I begin by providing a decision problem in which the probability–weight distinction has behavioural consequences and then consider the grounds for judging the behaviour rational or otherwise. The major part of this chapter is devoted to the construction of an account of rational choosing on which a principled judgement might be made.

We start with the following problem. Imagine an agent faced with a choice in which an experimenter tosses one of two coins, A and B. A is recently minted, to within fine tolerances, and has turned up heads on 25 out of 50 tosses.[4] About B, the agent is told nothing except that the coin has been minted within the past 500 years. The agent can select a coin to be tossed and call a side: \$10 is won if the coin tossed lands with the called face up.

According to Savage, we can infer for all *rational* agents a probability measure from responses to a sequence of pairwise counterfactual choices. Suppose that the agent has pairwise preferences of the form indicated in Table 8.1 and that the agent's declared choices are maximal. Then we can infer from the preferences the probabilities that also appear in the Table.[5]

---

[4] It has been suggested by a number of people that this might be suspicious. However, the *maximum likelihood* estimate would not be 0.5 for both sides if the proportions are changed. Null hypothesis testing would treat only relatively large differences as indicative of non-equally probable outcomes.

[5] It so happens that this is the modal preference type. However, the problem is symmetric and arises for uncertainty preferers also.

Table 8.1 Preferences and probabilities

Pair	Preference	Probability Inferred
(1)	$H_A P T_B$	$p(H_A) > p(T_B)$
(2)	$H_A P H_B$	$p(H_A) > p(H_B)$
(3)	$H_A I T_A$	$p(H_A) = p(T_A)$
(4)	$T_B I H_B$	$p(T_B) = p(H_B)$
(5)	$T_A P H_B$	$p(T_A) > p(H_B)$
(6)	$T_A P T_B$	$p(T_A) > p(T_B)$

H_A denotes heads on coin A and the terminology follows obviously. The six preferences exhaust the pairwise comparisons between the four possibilities. If heads and tails are exhaustive on coin A, the conjunction of (3), (2), and (6) forces the interrogator to infer that the subject has sub-additive probabilities.[6] (Other combinations of pairs generate problems, but for expositional reasons it is necessary to give only one triple which demonstrates the problem.) Alternatively, if it is assumed that heads and tails on B are exhaustive, (4), (2), and (6) yield super-additive probabilities on A. So we appear to have a situation in which subjective probability cannot possibly model behaviour motivated by concern for weight, and such behaviour we might call uncertainty-averse.

The example can also act as leverage against proposals concerning second-order probabilities. Assume the argument is that second-order probabilities may be used to describe uncertainty by collapsing down on the first-order ones through normal multiplication procedures. If the probabilities in final form are sub-additive, and if second-order probabilities collapse down to admissible final-form probabilities, then in unreduced form, at least one level of the probabilities would have to be sub-additive.

Were it possible to describe the differential uncertainty in terms of sets of permissible probability functions, if $\omega(p)$ really is

---

[6] Sub- and super-additive are here used to denote situations in the probability of the universal event being less than, or more than, one.

a probability, it would logically follow that the analyst should calculate a single figure by adding the products of the probabilities. So long as the final figure denoting the probability is used in decision calculations, any differences at the second-order level have no direct behavioural impact. But if we feel that ω is an argument of normative utility functions which can be sensitive to the way in which a collapsed probability, p_1 say, was originally constructed, then it seems that we cannot allow that ω is a second-order probability.

The Normative Issue

The inability of subjective probability to describe behaviour sensitive to the uncertainty-distinguishing options is but a starting-point. The central question to which the problem above leads is a normative one: namely, how agents *should* decide.

The issue can be explored in the light of Savage's comments about the role that axioms might play as logical policemen. If preference axioms were like rules of logical inference, it should be possible to find a change in the agent's choice behaviour which it is rational to make. There are a number of ways in which choice might be modified, but the most plausible seem to be the following:[7]

> STRATEGY 1: The agent alters the indifference relation in (4) to a pairwise preference relation of either sort, and modifies the preference relations in (6) and (2) appropriately.
>
> STRATEGY 2: The agent alters the preference relations in (1), (2), (5), and (6) to relations of indifference.

Both strategies solve the subjectivists' problem by resolving the 'inconsistencies' noted above. However, a consequence of changing (4) is that the subjectivist now infers that the agent believes $p(H_B) \gtrless p(T_B)$. Under the circumstances, this seems to be an unwarranted belief for the agent to hold, and it would seem

---

[7] Expositionally, it seems clearer to give only one set of changes for each strategy, though the reader will find that the symmetry of the problem helps to generate a number of other possibilities.

even stranger if our notion of rational choice forced such a preference pattern upon us. Strategy 2 makes life easier by allowing the subjectivist to infer that the agent assigns equal probabilities to each possibility. However, it does so at the cost of not being able to identify the fact that the agent assigns different weights to the probabilities. Selecting choices that indicate indifference where strict preference holds is a strictly dominated strategy and is not, therefore, obviously desirable. Implicitly, this argument draws on intuitions about the association of rationality with attitude-consistent behaviour, and though this seems to be central to optimization in choice, variants of SEU never formalize this aspect. In the following section I shall sketch this and other requirements that have claims as criteria that characterize rational choosing. Apart from any intrinsic interest the theory may offer, it will be apparent that such a theory would not offer any more support for Savage's interpretation and version of SEU than we have found thus far.

The intuitive argument may not be the most persuasive, but it helps outline the nature of the issues. Simply, it is far from clear that any epistemic measure, ω (measured, say, by n), is conceptually bounded from above in the way that p (the limit of r/n as n tends to infinity) is. Though probabilities of 0 and 1 are generally allowed, it is not obvious that we could ever attain complete certainty in the epistemic sense. Complete certainty about a world would presumably contain information about *all* the contents of that world. This, in turn, would include the set of all descriptions of the world which, to be *complete*, would contain a description of the complete description, a description of that description, and so on. This finite impossibility, which Popper (1950) named the 'Tristam Shandy paradox' in his analysis of quantum physics, suggests that, unlike $p(\cdot)$, $\omega(\cdot)$ cannot reach its upper bound.[8]

A second argument centres on the use of ω, which naturally

---

[8] Put another way, what is rational for infinite beings may not be even asymptotically rational for finite beings—see also Binmore (1987, 1988).

leads to a procedure for combining information that is not isomorphic with the use of Bayes's rule.[9] Consider a situation in which an agent samples a distribution, prior to which he holds the belief that the probability of an event, E, is 0.6. The sample consists of three trials and on one occasion E occurs. Using standard Bayesian reasoning, the agent calculates a new posterior probability estimate yielding results that are summarized in Table 8.2. On two occasions out of three, the agent observes $\neg E$, and a new posterior probability for E results.

Table 8.2 Weight versus Bayesian updating

Prior probability	Sample relative frequency	Likelihood	Posterior probability	p, ω
$p(E) = 0.6$	1	0.288	0.4	0.54, 13
$p(\neg E) = 0.4$	2	0.432	0.6	0.46, 13

Calculation of posterior probabilities from a sample size of three with prior probabilities as given is *independent* of the way in which the prior beliefs are formulated. However, suppose those priors happened to be based on observation of a different sample—say, one of size 10. It would be natural under these circumstances for an advocate of a two-dimensional approach to report, for each event, the ordered pair in the last column of the table. Not only is this consistent with maximum likelihood, but information about the epistemic state is conveyed which is never apparent either at the prior or the posterior stage under Bayesian procedures. In this case, a weight-based approach would combine the information sets to give a maximum likelihood estimate of probability and a measure of ω (based on the sum of the sample sizes, $10 + 3$). In general, prior and sample probabilities might be combined in proportions that reflect weights assigned by the agent. These weights need not

[9] The importance of the procedures for combining information can be seen in Arrow (1951), in which the system of potential surprise is rejected on the grounds that it has no such mechanism.

be limited to sample size but might reflect other epistemic concerns.[10]

No one would argue that any of the positions above are decisive, although they point towards Keynes and Ellsberg rather than towards the subjective probability view. In the next section I sketch axioms for rational choosing that provide a more substantial basis on which judgements can be made.

FOUR AXIOMS FOR RATIONAL CHOOSING

The theory outlined in this section draws on intuitions about the nature of rationality that many decision theorists, particularly philosophers, will recognize. My aims are to capture constructs that can be found in natural language and to show that they can be brought on to a linguistic par with those notions that economists presently think of as necessary conditions for 'consistency'.[11] It will become evident that the resulting theory of rationality looks rather different from those theories we traditionally associate with the foundations of decision theory, and much closer, for example, to theories of rationality such as that proposed by Sen (1985). A certain amount of methodological argument will help indicate why potential objections to the axioms proposed should not be considered successful or relevant.

Primitives

Any theory needs some minimal account of the primitive *objects* over which the primitive relations are to operate. Now while, *qua* decision-makers, we may be concerned ultimately about the status of ontological entities, it is important to accept that, conceptually, a decision theory must be defined on the primitives that it is logically feasible for a decision-maker to use; ontologi-

---

[10] Reading Fishburn (1983*b*), I think it might be possible to claim that this is consistent with Bayes's original theorem but not with modern extended use of that theorem.

[11] I shall not explicitly use set-theoretic notation, but the natural-language descriptions are readily formalizable.

cal entities will not do, though *finite conceptualizations* of those entities might. Perhaps signalling his awareness of the problem, Jeffrey (1965) lets propositions be primitive in his version of SEU and he defines utility of a proposition, the *news value*, as if it were true.[12]

An alternative, a sentence-based primitive, is discussed by Achinstein (1983) in his analysis of scientific explanation. In that context he finds various problems with a non-propositional foundation that points to the superiority of propositional primitives.[13] For present purposes I shall use a subset of propositions which I shall call *descriptive propositions*. In general they have the following form:

(D1): Option *a* has (*q* quantity of/more) property α (than option *b*).[14]

The atoms of the theory serve to attach properties to options. Associated with each is a probability, i.e. the probability that the proposition is true, and associated with each probability is an epistemic measure of the weight of evidence attaching to that probability. The building blocks of the theory can be thought of as a triple $t \equiv \langle d, p(d), \omega(p) \rangle$ with the ith choice option in decision calculations being some finite set of these triples T^i ($\equiv \bigcup t$). We can further define an information class $\mathfrak{z}$, as $\mathfrak{z} \equiv (\bigcup T^i : \forall T^i \in \mathbb{F}) \bigcup (\bigcup T^i : \forall T^i \in \mathbb{F}')$, where $\mathbb{F}$ is the class of feasible options and $\mathbb{F}'$ is the class of infeasible options. In principle, I think we *should* want a perfectly rational agent to act on the basis of $\mathfrak{z}$ and all the logical consequences of the conjunction of logical propositions. Unlike SEU, there is no requirement that option descriptions are materially complete—in reality, material completeness is a luxury rarely affordable.

---

[12] This proposal Jeffrey (1965) attributes to Savage.

[13] Also note that standard expected utility theory *à la* Savage avoids the problem, not by finding other primitives, but by assuming states independent of acts. The assumption is not part of the explicit axiomatization but needs to be made, an instance of what might be called a 'soft' axiom. See Malinvaud (1952) for an instance of soft axiom discovery in von Neumann and Morgenstern (1944).

[14] The properties can be constructed as predicates of first-order logic and the probabilities and weights can be incorporated into the predicates.

DOMINANCE

One of the fundamental insights that economics brings to the theory of rationality is the notion that one should not choose those things that are dominated. Indeed, so basic is the proposition that one has to take care not to let the requirement fall into the ravines of tautology. What economists today think of as dominance can be traced back to the pursuit of self-interest, which played a central role in early political–economic analyses. Yet for an idea that permeates the fabric of economics, there are some curiously disparate interpretations and understandings of what is being claimed by advocates of particular theories. In one of the higher profile exchanges, Boland (1982) claimed that utility maximization was an untestable, metaphysical proposition. Rather more conclusively, Sen (1985) points out that a person could be consistent in the axiomatic sense of SEU and yet choose in opposition to his or her preferences.

An alternative way of making Sen's point seems to be this. The problem with SEU is that it forgets to formalize (perhaps because this is of no significant mathematical interest) a principle that seems to drive, *inter alia*, Samuelson's (1938) revealed-preference theory: namely, that rational choices should reflect an agent's preferences. Assuming that it makes sense to think of a set of options, all of which are preferred to members outside the set but which cannot further be ranked in preference terms, as a maximally preferred set of options, a dominance axiom can be defined thus:

> AXIOM 1 (A1): *Dominance.* If the set of maximally preferred options is not empty, then the agent should choose an element from that set.

The relation of this requirement to rational choosing I believe to be self-evident and primitive. Most theorists would say the same, though there is probably less agreement about what forms of dominance are appropriate. I suspect that very few would object to the version above, in a sense that it summarizes the very rationale for normative decision theory, but it is fair to ask whether a distinction between behaviour and preference leads to propositions that have testable behavioural content.

Although convention does not require it, we may draw a distinction between behavioural content and testable content. As far as behavioural content is concerned, the axiom offers a non-tautological, conditional imperative concerning an agent's choice. What the person should do depends completely on his or her preferences, but once these are fixed, the set from which a choice should be made is completely identified.

Testability raises other issues. For one thing, it is evident that, even though the set of maximally preferred options may be determined, we might not know what it is. If we have no method of observing preferences independent of choices, then the contents of an agent's maximally preferred set must remain a mystery. No doubt, actions speak louder than words; but this seems, in some circles, to be equated with the rather extreme view that words are mute. In any case, even if we allow that the behavioural content of dominance is not testable, it is by no means obvious that this provides sufficient grounds for disregarding the axiom. First, we see that the open-ended scope for (and use of) primitive redescription makes SEU's own axioms untestable (see Chapter 7). Even if failure to satisfy testability is undesirable, axioms like transitivity are no better than dominance. Second, and more fundamentally, we may question just how important testability is. One need only consider for whom the lack of testability would be an embarrassment. To be sure, observers can use theories to make predictions only if those theories pick out proper subsets of what is feasible. But the major normative objective of decision theory is to specify how an idealized deliberator might choose. It is not clear what advantage testability is to a person trying to optimize—it might be added that randomizing strategies in games are rational precisely because the behaviour they produce is not testable.

To summarize, methodological objections to dominance may be rejected on the grounds that the axiom does have behavioural content, and that testability is a concern of the observer that should not be confounded with those of the decision-maker.

INTENTIONALITY

Cedric, a consumer, prefers doughnuts to bananas because they taste better. On a recent shopping trip he left the house meaning to get some doughnuts but forgot the aim of his visit when he arrived at the supermarket. However, during his shopping spree he brushed past a bag of doughnuts which fell into his basket of goods, a fact that he discovered only after he had paid for all the items. While it would be true to say that Cedric *bought* doughnuts, and not bananas, and that the selection satisfied (A1) because doughnuts were strictly dominating, it would seem a misuse to call purchase behaviour a *choice*, let alone a rational choice. Choice seems to presuppose some sort of causal mental state, though not all mental states will do. Cedric might be prone to fits induced by shopping-mall muzak. It may be that during one of these fits he shakes in such a way as to accidentally knock the doughnuts into his basket. Here we have a causal mental state which does not make the behaviour deliberate, and we need to eliminate such cases. Joining these considerations, I shall require a behaviour to be deliberate in the following sense:

AXIOM 2.1: *Deliberateness.* There exists a causal[15] mental state and it is an 'in control' mental state.

What counts as an 'in control' mental state varies between legal systems, societies, and schools of psychology, so it is not possible to give universal criteria. However, extremes of grief, provocation, and drug abuse are examples of situations in which, for our society, the existence of causal mental states do not guarantee deliberateness. It might be objected that the theory should be more definite about the nature of primitives such as 'in control' mental states, but in defence, it should be noted that standard theories also invoke constructs that are left incompletely explicated. It is hardly a settled matter as to whether, for instance, a preference is a mental state, a set of mental states, a proneness to having a mental state, or something else.

---

[15] Here I presume some non-evidential account of causality such as that used in Fodor (1968).

A consequence of Davidson's (1963) argument seems to be that deliberateness cannot be equated with intentionality. You may buy (deliberately) a second-hand car and have it turn out to be a lemon (unintentionally), to borrow an example from Akerlof (1970). Similarly, it was a deliberate act that caused Oedipus to marry his mother, though he did not intend to do so. The philosophical conclusion, i.e. that intentions are intentional (Anthony 1987), gives rise to a further requirement, which I shall call 'intentionality', and define as follows:

> AXIOM 2.2: *Intentionality.* The set of things the agent knows contains at least one description under which the choice was intended.

Jointly, (A2.1) and (A2.2) are designed to capture the intendedness of a behaviour. Classical decision theory is not explicit about this concept. However, we might note that, if unintentional behaviour is preference-revealing only by chance, it is reasonable to suggest that classical consumer theory assumes intentionality implicitly (Federkke (1993) provides a very substantial new analysis of Davidson and the role of intentions in explanations of rational actions).

INSTRUMENTALITY (KNOWN PREFERENCE-DRIVEN CHOICE)

If we are to take seriously the proposal that rational choice is not merely a consequential notion, we have to face up to the possibility that we might get the right answer for the wrong reasons. Maher (1986), for instance, notes that 'for a rational choice the person must not only choose the best act, but must also choose it *because* it is the best act'. To reflect the importance of appropriate causality, a further condition is proposed:

> AXIOM 3: *Known preference-driven choice.* The causal mental state is preference-motivated.

The principal role of this axiom is to exclude the possibility that behavioural intentions, formed by random processes like coin tossing instead of being determined by relevant preferences, lead to rational choice. They may lead to maximally preferred selections, as game-theorists have pointed out, but that is something else. It may be that, like the dice man, the agent eschews rational choosing as a method of selection, and perhaps for good dominance-based reasons. Under such a situation, there is no difficulty in saying that the method of choosing was itself the subject of a rational choice, but that any options selected by this method are not themselves rationally chosen.

PRUDENCE—HAVING A REASON

It may seem strange that most economic theories of *rational* choice make no explicit reference to the notion of *reason*. Chambers's dictionary includes in the definition of reason, 'ground, support, justification of an act or belief, motive or inducement, underlying explanatory principle, a cause, and conformity to what is fairly to be expected'. These aspects of 'reason' share a common element which can, and should, be included in a full account of 'rational choice'.

To see the significance of reason, consider a situation in which dominance is satisfied but no ground for the choice exists. It might be claimed that dominance provides a sufficient, if brute, reason for that choice. Theories of rational decision-making may just be an elaboration of what dominance requires in certain situations. However, this position is not wholly satisfactory in that it allows, *inter alia*, the capricious and whimsical to justify every possible choice act on the sole basis that it is the preferred act. Such justifications are minimally informative and not in any normal sense prudential. Behaviours that satisfy dominance may be the result of maximizing actions, but without further warrant they are *ad hoc*.

For a choice to be rational, I shall require that there exists at least one reason supporting the choice, where a reason for a choice a, $r(a)$, can be characterized as a pair of propositions of the following form:

(P1): *More/less of property α is desirable.*
(P2): Option a has more/less property α than option b

Note that the second proposition is an instance of the descriptive proposition (D1) in its *comparative* form. Using this definition, the fourth and final axiom can be described.

>   AXIOM 4: *Minimal prudence.* The class of reasons, $R \equiv \{r(a)\}$, for choosing an option a, must not be null and the class of things the agent knows must contain at least one element of R.

A situation where an agent can explicitly rank options according to attributes would count as one where reasons clearly exist and are known to the decision-maker. This requirement is not imposed by SEU, nor is it innocuous, for it could be taken that the chosen option has a property shared with one of the options rejected. Consider, for instance, a group of friends deciding whether to go out for a meal or to watch a film; should they be allowed to say that they chose the meal over the film on the grounds that the meal was likely to have a better flavour? Some philosophers take the view that attributing a zero level of a predicate to an entity that would not in any normal use of the natural language be said to have that predicate is legitimate. However, there are others who think that such a move is not acceptable. Intuitively, the latter position seems less strained; besides, it gives more content to the notion of having a reason.

Sufficiency of Reason

An important issue in any discussion of reason-having is the question of whether a reason is a good one or not. Is the reason *sufficient* to warrant the acts that are said to follow from it? Kane's (1986) analysis of the logic of sufficiency of reason shows how extensive the concept is, but not how we might measure sufficiency or sort good reasons from bad ones. The question looks difficult until we recognize that there are essentially three different systems for determining sufficiency.

In the first instance, it is possible to regard the desirabilities that constitute reasons as depending on moral judgements of desirability. As properties of action will not be equally desirable under different moral systems, the question as to whether there is a reason for a choice can only be given a system-relative answer.

The second approach is what I shall label *utilitarianism*, and it has two basic forms. In the first, desirabilities are determined solely by the agent's preferences. It may be impossible to make the neuro-physiological measurements required to determine directly what the agents' preferences are (assuming that they are indeed brain states), but the question is, in principle, completely empirical. This form, which I take to be consistent with economic theorizing in general, and which most economists use, perhaps implicitly, I shall call an *extreme psychologistic view*.[16]

There is a second brand of utilitarianism that can be identified. This regards utility as some form of objective need which may not always be realized as preference. This view is no longer fashionable among economic theorists, though something like it seems to inform the work of political economists up to the middle of the last century, and of development specialists and industrial psychologists.

Last but not least is what might be termed the *conventional* or *anthropological account of sufficiency*, which derives from nothing more or less than cultural expectations and notions of what is a good reason. For instance, who has not at some time felt the urge to buy a book simply because it has an attractive cover and wondered whether this was really a sufficient reason for purchasing the book? The law makes much of this sort of sufficiency (for instance, what counts as negligence). In so far as none of us comes to hold the values we do *in vacuo* (cf. John Donne's 'No Man Is An Island'), this third sense is not wholly separable from that embodied in the psychologistic brand of utilitarianism.

Four axioms—dominance, intentionality, known preference-driven intentions, and prudence—sketch intuitions about rational choosing. A case for holding to these requirements has been made, and if the primitives and axioms are accepted, then behaviour

---

[16] A similar term appears in Coats (1988). It is this view that is consistent with Ernst Mach's view about physics: 'why for example, only weights and distances, and not, say, colours and shapes, are relevant to the lever's motion. No *a priori* principles tell me that' (Kane 1986: 119).

outside SEU cannot automatically be judged irrational. A person who chose deliberately and intentionally, had preference-driven intentions, and gave different degrees of uncertainty as a reason for choosing would be making a rational choice by violating SEU.

FINAL REMARKS

The specific thesis of the chapter is that sensitivity to uncertainty, measured by the weight of evidence and not describable by SEU, can be rational. The theory of rationality proposed in order to make the judgement draws on philosophical intuitions about the nature of rational choice but is not without some support in economics. Sen (1985), for instance, takes a critical view of traditional consistency axioms and proposes two new kinds of consistency requirements:

> CONDITION 1: *Correspondence*. A person can fail to do what he would decide to do if he were to reason and reflect what is to be done.

> CONDITION 2: *Reflection*. A person may have reflected as carefully as he can on a choice, but not have seen something significant that sharper reasoning would have revealed.

Sen argues that the criteria embedded in these conditions better describe the consistency that rationality requires than do the standard axioms. Of course, there are differences between Sen's theory and that presented here; but it should be evident that neither leads one to suppose that sensitivity to uncertainty is, a priori, irrational. My strategy has been to provide an account of rationality based solely on the evidence of competent users of the natural language. No attempt was made to constrain the answer to have particular methodological or substantive properties, and the result is a theory that represents a marked departure from SEU.

9

Summary and Conclusions

Although other models are useful, it is the interpretation that
SEU explicates ideally rational decision-making that intrigues
us. In psychology it provides a basis on which rationality can be
tested; in economics it specifies the sense of rational agency on
which much analysis is based; in statistics it provides support
for the subjective approach that Bayesians advocate; in philosophy it contributes not just a new account, but a new, formal
type of rational choice theory. Of classical decision theory itself,
we might say that SEU is its heart and soul.

There can be few theories that would appear to be so important in so many disciplines. No one would say that SEU's days
are over, but equally, the pressure for a substantial re-evaluation
of its status is becoming harder to resist. The numerous and
systematic experimental falsifications of SEU's axioms are now
broadly accepted. The existence of a wider class of non-expected
mathematical theories is now well known. If there ever was a
kind of infant industry argument[1] for insulating SEU from
criticism, those days are long gone. As Levi (1986) suggests, the
time for critical inquiry is upon us, though the only non-pragmatic argument yet to be resolved concerns SEU's normative
model.

It should come as no surprise to learn that normative issues
are the last to be resolved. The argument is conceptual and
therefore depends on more than just the decisive techniques that
logic, mathematics, and even experimentation offer. The resolution of such issues is rarely as clean as for technical ones, but it

---

[1] Popper came to the view that dogmatism could override rational criticism in
the early stages of a theory's life. The parallel with the argument for protecting
young industries from competition suggests the danger that theories could
become protection-dependent.

would be a mistake to think that it is, therefore, an essentially arbitrary enterprise. What we can say I summarize below, concluding with remarks about future work and methodological implications.

PREFERENCES

Of the axioms that Savage uses to ground SEU, we examined the warrants for those that have been given substantive interpretations. These assumptions are transitivity, independence, completeness, and continuity, although I have concentrated, as the literature does, on the first two of these. Neither is necessary for mathematical–economic analysis anymore. There may be more evidence against independence, though the problems with the arguments for transitive preference are at least as considerable as those for independence. On many occasions, the warrants for both axioms appear to fail for similar reasons, which suggests that some of the arguments may be less distinct than would at first appear.

We should note that there *are* features of decision problems that can be used to explain why people want to violate SEU. This does not prove that such behaviour is rational, but it establishes the possibility that such a demonstration might be given. The motive for doubting the equation of SEU with rationality comes from a logical analysis of the arguments that accompany its assumptions. Closer inspection reveals that the normative interpretation of SEU depends on incomplete arguments, implicit assumptions for which a normative case has yet to be made, analogies that are question-begging, or the confounding of constructs that are reasonably distinguished. There are, moreover, situations in which optimization seems to mandate intransitive choice. Attempts to cut the ground from under such examples exist, but they depend on redefining primitives in ways that are either *ad hoc* or principled but behaviourally untestable.

BELIEFS

Like knowledge, uncertainty shapes the facts of economic and social life. Aspiration, fear, hope, surprise, trial, learning, forecasting, negotiation, insurance, and even competition all depend in some measure on the future not being known for certain. Economists are right to emphasize the importance of uncertainty, and for the most part it is the calculus of probability that is used to make intelligible the issues it raises. The frequency, logical, and subjective models of the calculus provide accounts of what probability is and how best it is measured. Less directly, they suggest ways in which the purely formal calculus might be useful, though no single account predominates and each is known to have flaws. It is against this backdrop of usefulness and imperfection that the developing inquiries into the modelling of rational credence should be seen.

The history of *extra*-probabilistic analyses of belief is older than the view that subjective probability does what relative frequency cannot. The older position has never attained widespread acceptance, but neither have the rejections been persuasive enough to eliminate lingering doubts or prevent the re-emergence, from time to time, of proposals for non-probabilistic concepts of credence.

The view that epistemic concerns are real but are not the same as probabilistic ones is not as marginal as we may think. Classical hypothesis-testing, for instance, does not rely solely on comparing maximum-likelihood estimates of parameters. Admittedly, the significance levels used are arbitrary, but the principle that the decision to reject an hypothesis is a function of the quantity of information on which estimates are based seems to be a good one and one we should not reject.

It will be countered that Bayesianism offers many things that classical statistics do not. The evidence of theory can be recognized through the use of prior probabilities; these can be combined with further evidence; and in any case, Bayes's theorem is a mathematically correct account of the relations between conditional probabilities.

The remark that we need a set of probabilities *and* a measure of information quantity to describe completely the distributions that we actually observe would be trivial if practice were not otherwise. In any case, the implications of adopting a two-dimensional theory of probability are rather more substantial. The main issues seem to be as follows. First, we can show how uncertainty-averse behaviour is compatible with the holding of probabilities that sum to unity and is not, therefore, susceptible to Dutch-Book-based charges of irrationality. Second, it is possible to pool prior and other probabilities using a weighted average where the weights reflect quantities of information on which each probability is based. This represents an improvement on Bayesianism in the sense that it allows for information-pooling which is consistent with the maximum-likelihood principle. To many statisticians, maximum likelihood has the status of an axiom and is less questionable than the Bayesian procedures that violate it. Third, we can let 'prior' probabilities be accorded different weights—the Bayesian approach forces us to treat all priors as equal even if we know that some are based on more evidence or better theory than others.

DIRECTIONS FOR THE FUTURE

If rationality cannot be associated with the assumptions of SEU, one may reasonably ask what rationality is. The account sketched in Chapter 8 suggests that dominance, prudential reasoning, and intentionality play a role. Possibly the last two of these are of more importance in philosophy, but few could argue about the necessity of dominance, though the characterization of it as a relation between preference and behaviour is, perhaps, contestable. Preferences may be (even) harder to observe than behaviour, but it seems unhelpful to assume that they are therefore non-existent. Some economists are, no doubt, philosophical behaviouralists, but this reduction of existential issues to epistemic ones leaves the motivation for normative inquiries obscure.

The account of rational choosing proposed in Chapter 8 offers little prospect of *deriving* unconditional constraints on behaviour, but I do not think it is the primary role of a subjective theory of rational choice to do this. Perhaps we should regard more general theories as applying to increasingly sophisticated decision problems. Ultimately, I see only violations of dominance (in the sense of Chapter 8) as irrational. Under very few circumstances, this will lead to rejection of (1) logical inference, (2) causal utility (for the evidential model), and (3) the interpretation of Bayes's theorem as relating known conditional probabilities (but not the application to updating beliefs). Those who want to violate expected value should be asked if it makes sense to take account of value in a non-linear way. If risks are large compared with current wealth, it seems reasonable to use utilities instead of values. Similarly, large risks might justify weighing probabilities in a non-linear way. If agents want yet further freedom to act, we may ask, for instance, whether they think it reasonable to take into account counterfactual possibilities. If they do, then we need to move to an even wider class of models that drop Hammond's consequentialism. In other words, there is a core of things that rational agents should not violate. Beyond that is a hierarchy of increasingly general models. None of the models is in any absolute sense a model of rational choice, but if the agent wants to proceed to a more general model, the step up to a more general class of theory provides an opportunity to reflect whether this is really what he or she wants. Ultimately, there is no limit to the models that might be used.

The relationship between mathematical theory and experimental evidence makes decision theory one of the most physics-like parts of economics and appears, at present, to be productive. I believe that the mutual gains for theorists and experimentalists will ensure that the area flourishes, though like Peter Fishburn I see axioms becoming less important. For one thing, if, as experimental evidence and intuition suggests, preferences are stochastic, then any test of a deterministic axiom will result in errors. Further, any investigation of a different partitions of the error set will eventually find some in which the errors are systematic. I hypothesize, therefore, that ultimately the current pattern of deterministic axiom evaluation by systematic error finding will collapse. On the other hand, greater emphasis on the form of the utility function may also lead to greater interest in multi-attribute utility functions, which are surely more relevant for most decisions. This may in turn lead to an increasing reliance on stable empirical relations to obtain behavioural content, and I should like to see this come, not just from experiments, but from ethnographic field studies. Decision theory may, as a result, take on new significance for game theory. Rather than providing foundations, it may offer solution concepts for the majority of games that have no 'analytical' solution.

What distinguishes mathematical economic theory from pure mathematics is its interpretation. If views on the rational status of SEU shift, the change will constitute a conceptual volte-face of immense proportions. The lesson to be learnt is not that mathematical modelling should be abandoned, but that it is a fallible driver of intellectual progress. If axiomatic theories are fallible, then surely no method of inquiry is privileged, and pluralism seems the only defensible methodology. One need not disagree with Hilary Putnam's view that mathematics needs no foundations to suggest that areas of applicable mathematics such as physics and economics do, and that philosophy provides ways of thinking useful to the task. Indeed, the mark of economics' maturity might seem to be not the fact that it increasingly uses mathematics as a source of innovation rather than a 'mere' formalization, but rather that the mathematics it employs is beginning to raise 'philosophical' questions that are on a par with those raised by physics and even by pure mathematics itself.

Further Reading

Typically, decision theorists specialize in one of the areas below and keep abreast of some of the others. Many social scientists and philosophers concentrate on those parts of decision theory that correspond most closely to the methods of their own discipline; and, while this is understandable, it means that one rarely gets an overview of the subject in its entirety. At the risk of giving offence to the authors of many excellent texts and collections of readings, I offer here a baker's dozen of books that the student, interested in obtaining a cross-disciplinary understanding of decision theory, might want to consult.

NORMATIVE THEORY

The book that deals with many of the important philosophical issues is the reader *Decision, Probability and Utility*, edited by Gardenfors and Sahlin (1988). Whether all uncertainty can be handled within a probabilistic frame is an issue that arises indirectly in Part III and directly in Part IV. In my view, the collection of papers they assemble highlights rather clearly some of the key reasons why the possibility must be rejected. In a similar though broader vein, *Foundations of Decision Theory*, edited by Bacharach and Hurley (1991), provides a survey of issues in largely original papers by philosophers and economists brought together with a particularly strong introduction by the editors. Finally, I mention three books that are of more methodological relevance. *Beyond Positivism* by Caldwell (1982) and *The Inexact and Separate Science of Economics* by Hausman (1992) offer excellent discussions of economic methodology, and both have chapters that deal specifically with rational choice theory. *Mathematics, Matter and Method* by Putnam (1979) is as good a starting-point as any if you want to explore parallels between mathematical method in geometry or physics and axiomatic decision theory.

MATHEMATICS

A classic book that focuses on the mathematics of non-expected utility is *Non-linear Preference and Utility Theory* by Fishburn (1988a) and no doubt this will continue to be a standard reference for some years to come. Non-mathematicians might benefit from reading *Games and Decisions* by Luce and Raiffa (1957). Although the contents reflect the age of the book, this is probably still one of the best and intuitively insightful introductions to decision theory.

EXPERIMENTAL DECISION THEORY

One of the more comprehensive introductions to the experimental literature (which is vast) can be found in *Thinking and Deciding* by Baron (1988). While the book is aimed at psychology undergraduates, there are a number of chapters that would appeal to anyone interested in experimental economics. A rather useful, though less extensive, alternative is the earlier *Human Inference* by Nisbett and Ross (1980).

DECISION ANALYSIS

As one might expect, decision theory has an applied side to it. The way decision theory helps decision-makers is not nearly so crude as some people might suppose, and *Decision Analysis and Behavioural Research* by von Winterfeldt and Edwards (1986) makes the point as well as any such text.

APPLICATIONS

It is only a theorist's bias that causes me to put the applications last, for in reality one can learn a lot about theories by applying them (or trying to). *Agricultural Decision Making* edited by Barlett (1980) includes papers by a number of anthropologists, ranging from applications of Tversky's elimination by aspects model to theories of lay descriptions of uncertainty. Medicine is another area where decision theory has been applied extensively, and a number of medical decision-making issues are discussed in *Professional Judgement* edited by Dowie and Elstein (1988).

A GENERAL READER

There are a number of general collections of readings, and one that I think has a nice mix of theoretical perspectives as well as some applications is *Decision Making* edited by Bell *et al.* (1988).

Bibliography
(*includes works not specifically cited*)

ACHINSTEIN, P. (1983), *The Nature of Explanation*, Oxford University Press.

AKERLOF, G. A. (1970), 'The Market for "Lemons"', *Quarterly Journal of Economics*, 84: 488–500.

—— and DICKENS, W. T. (1982), 'The Economic Consequences of Cognitive Dissonance', *American Economic Review*, 82: 307–19.

ALCHIAN, A. (1953), 'The Meaning of Utility Measurement', *American Economic Review*, 43: 26–50.

ALDERFER, C. P. (1969), 'An Empirical Test of a New Theory of Human Needs', *Organisational Behavior and Human Performance*, 4: 142–75.

ALLAIS, M. (1953), 'Le Comportment de l'homme rationale devant le risque, critiques des postulates et axiomes de l'école Americaine', *Econometrica*, 21: 503–46.

ANAND, P. (1993) 'The Philosophy of Intransitive Preference', *Economic Journal*, 103: 337–46.

—— (1991), 'The Nature of Rational Choice and the *Foundations of Statistics*', *Oxford Economic Papers*, 43: 199–216.

—— (1990c), 'Analysis of Uncertainty as Opposed to Risk', *Agricultural Economics*, 4: 145–63.

—— (1990b), 'Two Types of Utility', *Greek Economic Review*, 12: 58–74.

—— (1990a), 'Interpreting Axiomatic (Decision) Theory', *Annals of Operations Research*, 23: 91–101.

—— (1987), 'Are the Preference Axioms Really Rational?', *Theory and Decision*, 23: 189–214.

—— (1986), *Axiomatic Choice Theory*, D.Phil. thesis, Bodleian Library, Oxford.

—— (1982), 'How To Be Right without Being Rational', *Oxford Agrarian Studies*, 11: 158–72.

ANTHONY, L. (1987), 'Attribution of Intentional Action', *Philosophical Studies*, 46: 311–23.

ARROW, K. J. (1982), 'Risk Perception in Psychology and Economics', *Economic Enquiry*, 20: 1–9.

—— (1971), *Essays in the Theory of Risk Bearing*, Markham, Chicago.

—— (1951), 'Alternative Approaches to the Theory of Choice in Risk-Taking Situations', *Econometrica*, 19: 404–37.

—— and HURWICZ, L. (1972), 'An Optimality Criterion for Decision-Making Under Ignorance', in C. Carter and J. L. Ford (eds.), *Uncertainty and Expectations in Economics.*

AUMANN, R. J. (1962), 'Utility Theory without the Completeness Axiom', *Econometrica*, 30: 445–62.

BACHARACH, M. (1991), 'Games with Concept-Sensitive Strategy Spaces', paper presented to the International Conference on Game Theory, Florence, June.

—— (1990), 'Commodities, Language and Desire', *Journal of Philosophy*, 87: 346–68.

—— (1989), 'Zero Sum Games', in J. Eatwell, M. Milgate, and P. Newman (eds.), *Game Theory*, Macmillan, London.

—— (1986), 'Rational Preferences and the Agent's Descriptions', mimeo, Christ Church College, Oxford.

—— and HURLEY, S. L. (eds.) (1991), *Foundations of Decision Theory: Issues and Advances*, Oxford, Blackwell.

BARBERA, S., BARRETT, C. R., and PATTANAIK, P. K. (1986), 'On Some Axioms for Ranking Sets of Alternatives', *Journal of Economic Theory*, 32: 301–8.

BARBERA, S., and PATTANAIK, D. K. (1986), 'Extending an Order on a Set to the Power Set', *Journal of Economic Theory*, 32: 185–91.

BAR-HILLEL, M., and MARGALIT, A. (1988), 'How Vicious are Cycles of Intransitive Choice?', *Theory and Decision*, 24: 119–45.

BARNARD, G. A. (1961), 'Discussion', *Journal of the Royal Statistical Society*, 23: 25–7.

BARON, J. (1988), *Thinking and Deciding*, Cambridge University Press.

BARTLETT, PEGGY F. (ed.) (1980), *Agricultural Decision-Making*, London, Academic Press.

BAUSOR, R. (1988), 'Human Adaptability and Economic Surprise', in P. E. Earl (ed.), *Psychological Economics*, Kluwer Press, Lancaster.

—— (1985), 'The Limits of Rationality', *Social Concept*, 2: 66–83.

BECKER, S. W., and BROWNSON, F. O. (1964), 'What Price Ambiguity?' *Journal of Political Economy*, 72: 62–73.

BELL, D. (1982), 'Regret in Decision-Making under Uncertainty', *Operations Research*, 20: 961–81.

BEM, D. J. (1972), 'Self-Perception', in *Advances in Experimental Social Psychology*, Academic Press, New York.

BERNOULLI, D. (1739), 'Specimen Theoriae Novae de Mensura Sortis', in *Commentarii Academiae Scientiarum Imperialis Petropolitanae*, 5: 175–92 (English translation in *Econometrica*, 22 (1954): 23–36).

BEWLEY, T. F. (1986), 'Knightian Decision Theory: 1', Cowles Foundation Discussion Paper no. 807.

BINMORE, K. (1988), 'Modelling Rational Players: II', *Economics and Philosophy*, 4: 9–55.

—— (1987), 'Modelling Rational Players: I', *Economics and Philosophy*, 3: 179–214.

BELL, DAVID E., RAIFFA, HOWARD, and TVERSKY, A. (eds.) (1988) *Decision-Making*, Cambridge University Press.

BLYTH, C. (1972), 'Some Probability Paradoxes in Choice from among Random Alternatives', *Journal of the American Statistical Association*, 67: 366–73.

BOLAND, L. (1982), 'On the Futility of Criticizing the Neo-Classical Maximisation Hypothesis', 71: 1031–6.

BORCH, K. H. (1968), *The Economics of Uncertainty*, Princeton University Press.

BOSTIC, R., HERRNSTEIN, R. J., and LUCE, R. D. (1988), 'The Effect on the Preference Reversal Phenomenon of Using Choice Indifference', Working Paper, Harvard University.

BRADBURY, H., and ROSS, K. (1990), 'The Effects of Novelty and Choice Materials on the Intransitivity of Preferences of Children and Adults', *Annals of Operations Research*, 23: 141–59.

BRENNER, R., with BRENNER, G. A. (1990), *Gambling and Speculation*, Cambridge University Press.

BROOME, J. (1991), *Weighing Goods*, Basil Blackwell, Oxford.

—— (1989), 'An Economic Newcomb Problem', *Analysis*, 49: 220–2.

—— (1984), 'Rationality and the Sure-Thing Principle', Discussion Paper, University of Bristol.

CALDWELL, BRUCE (1982), *Beyond Positivism*, Allen and Unwin, London.

CHEW, S. H., and EPSTEIN, L. G. (1989), 'A Unifying Approach to Axiomatic Non-Expected Utility Theories', *Journal of Economic Theory*, 49: 207–40.

—— KARNI, E., and SAFRA, Z. (1987), 'Risk Aversion in the Theory of Expected Utility with Risk-Dependent Probabilities', *Journal of Economic Theory*, 42: 370–81.

COATS, A. W. (1988), 'Economics and Psychology', pp. 211–25 in P. Earl (ed.), *Psychological Economics*, Kluwer Press, Lancaster.

COHEN, J. L. (1985), 'Twelve Questions About Keynes's Concept of Weight', *British Journal of the Philosophy of Science*, 37: 263–78.

—— (1960), *Chance, Luck and Skill*, Penguin, Harmondsworth.

COHEN, M., and JAFFRAY, J. Y. (1985), 'Decision Making in a Case of Mixed Uncertainty', *Journal of Mathematical Psychology*, 29: 428–42.

—— —— (1980), 'Rational Behaviour under Complete Ignorance', *Econometrica*, 48: 1281–99.

CRAMER, G. (1728), Letter to Nicolas Bernoulli; cited in Bernoulli (1739).

CROON, M. A. (1984), 'The Axiomatisation of Additive Difference Models for Preference Judgments', pp. 193–207 in E. Degreef and B. von Buggenhaut (eds.), *Trends in Mathematical Psychology*, North-Holland, Amsterdam.

CROSS, J. G. (1983), *A Theory of Adaptive Economic Behaviour*, Cambridge University Press.

CUBITT, R. (1986), 'Two Perspectives on the Time Inconsistency Problem', *Greek Economic Review*, 8: 1–20.

DAVIDSON, D. (1963), 'Actions, Reasons and Causes', *Journal of Philosophy*, 60: 655–700.

—— (1980), *Essays on Actions and Events*, Clarendon Press, Oxford.

DEATON, A., and MUELLBAUER, J. (1980), *Economics and Consumer Behaviour*, Cambridge University Press.

DEBREU, G. (1959), *Theory of Value*, John Wiley, New York, for Cowles Foundation.

DOUGLAS, M., and ISHERWOOD, B. (1980), *The World of Goods*, Penguin, Harmondsworth.

DOWIE, JACK, and ELSTEIN, ARTHUR, S. (eds.) (1988) *Professional Judgement*, Cambridge University Press.

DRÈZE, J. (1974), 'Axiomatic Theories of Choice, Cardinal Utility and Subjective Probability', in *Allocation Under Uncertainty*, Macmillan, London.

DUBOIS, D. (undated), 'Possibility Theory', mimeo, Université Paul Sabatier, Toulouse Cedex, France.

EARL, P. (1983), *The Economic Imagination*, Wheatsheaf, Brighton.

EINHORN, H. J., and HOGARTH, R. M. (1986), 'Decision Making under Ambiguity', *Journal of Business*, 59: 225–78.

ELLSBERG, D. (1961), 'Risk, Ambiguity and the Savage Axioms', *Quarterly Journal of Economics*, 75: 643–69.

—— (1963), 'Reply', *Quarterly Journal of Economics*, 77: 336–42.

ELSTER, J. (1983), *Sour Grapes*, Cambridge University Press.

EVANS, J. ST B. T. (1982), *The Psychology of Deductive Reasoning*, Routledge & Kegan Paul, London.

FEDERKKE, J. (1983), 'The Use of Reason', Ph.D. thesis, University Library, Cambridge.

FESTINGER, L. (1951), *A Theory of Cognitive Dissonance*, Stanford University Press, California.

FEYNMAN, R. P. (1987), 'Negative Probability' *in Quantum Implications*, ed. B. J. Hiley and F. D. Peat, Routledge, London.

DE FINETTI, B. (1937), 'La Prevision: ses lois logiques, ses sources subjectives', *Annales de l'Institut Henri Poincaré*, 7: 1–68.

FISHBURN, P. C. (1988a), *Nonlinear Preference and Utility Theory*, Johns Hopkins University Press, Baltimore.

—— (1988b), 'Context-Dependent Choice with Non-Linear and Non-Transitive Preferences', *Econometrica*, 56: 1221–39.

—— (1983a), 'Ellsberg Revisted', *Annals of Statistics*, 11: 1047–59.

—— (1983b), 'Research in Decision Theory', *Mathematical Social Sciences*, 5: 129–48.

—— (1982), 'Non-Transitive Measurable Utility', *Journal of Mathematical Psychology*, 26: 31–67.

—— (1981), 'Subjective Expected Utility', *Theory and Decision*, 13: 139–99.

—— (1971), 'A Study of Lexicographic Expected Utility', *Management Science*, 17: 672–8.

FODOR, J. A. (1968), *Psychological Explanation*, Random House, New York.

FREGE, G. (1893), *Grundgesetze der Arithmetik*, i and ii, Jena, Pohle.

FRIEDMAN, M. (1953), *Essays in Positive Economics*, University of Chicago Press.

—— and SAVAGE, L. J. (1948), 'The Utility Analysis of Choices Involving Risk', *Journal of Political Economy*, 56: 279–304.

GAERTNER, W., PATTANAIK, P. K., and SUZUMURA, K. (1988), 'Individual Rights Revisited', mimeo, Department of Economics, University of Birmingham.

GALE, D., and MAS-COLLEL, A. (1975), 'An Equilibrium Existence Theorem for a General Model without Ordered Preferences', *Journal of Mathematical Economics*, 2: 9–15.

GARDENFORS, P., and SAHLIN, N.-E. (1982), 'Unreliable Probabilities, Risk Taking and Decision Making', *Synthese*, 53: 361–86.

—— —— (1988) *Decision, Probability and Utility*, Cambridge University Press.

GAUTIER, D. (1986), *Morals by Agreement*, Clarendon Press, Oxford.

GILBOA, I., and SMEIDLER, D. (1989), 'Maximin Expected Utility with Non-Unique Prior', *Journal of Mathematical Economics*, 18: 141–53.

GLADWIN, C. (1980), 'A Theory of Real Life Choice', in P. F. Barlett (ed.), *Agricultural Decision Making*, Academic Press, Orlando, Florida.

GOLDSMITH, R. W., and SAHLIN, N. E. (undated), 'The Role of Second-Order Probabilities in Decision Making', mimeo, Department of Psychology, University of Lund, Sweden.

GOLDSTEIN, W. M., and EINHORN, H. J. (1987), 'Expression Theory and the Preference Reversal Phenomena', *Psychological Review*, 94: 236–54.

GREEN, H. A. (1971), *Consumer Theory*, Penguin, Harmondsworth.

GREEN, J. (1987), 'Making Book Against Oneself: The Independence Axiom and Nonlinear Utility Theory', *Quarterly Journal of Economics*, 102: 785–96.

GRETHER, D. M., and PLOTT, C. R. (1979), 'Economic Theory of Choice and the Preference Reversal Phenomenon', *American Economic Review*, 69: 623–38.

GRIGGS, R. A., and COX, J. R. (1982), 'The Elusive Thematic-Materials Effect in Wason's Selection Task', *British Journal of Psychology*, 73: 407–20.

GRISWOLD, B. J., and LUCE, R. D. (1962), 'Choices among Uncertain Outcomes', *American Journal of Psychology*, 75: 35–44.

HAGEN, O. (1979), 'Towards a Positive Theory of Preferences under Risk', in M. Allais and O. Hagen (eds.), *Expected Utility Hypotheses and the Allais Paradox*, Reidel, Dordrecht.

HAKIM, C. (1987), *Research Design*, Unwin Hyman, London.

HAINES, G. H., and RATCHFORD, B. T. (1987), 'A Theory of How Intransitive Consumers Make Decisions, *Journal of Economic Psychology*, 8: 273–98.

HAMMOND, P. J. (1988a), 'Consequentialist Foundations for Expected Utility', *Theory and Decision*, 25: 25–78.

—— (1988b), 'Consequentialism and the Independence Axiom', pp. 503–16 in B. Munier (ed.), *Risk Decision and Rationality*, Reidel, Dordrecht.

—— (1976), 'Changing Tastes and Coherent Dynamic Choice', *Review of Economic Studies*, 43: 159–73.

HANDA, J. (1977), 'Risk, Probabilities and a New Theory of Cardinal Utility', *Journal of Political Economy*, 85: 97–122.

HANSSON, B. (1975), 'The Appropriateness of the Expected Utility Model', *Erkenntnis*, 9: 175–93.

HARSANYI, J. (1955), 'Cardinal Welfare, Individualistic Ethics and Interpersonal Comparisons of Utility', *Journal of Political Economy*, 63: 309–21.

HAUSMAN, D. M. (1988), 'Economic Methodology and Philosophy of Science', pp. 88–116 in G. C. Winston and R. F. Teichgraeber III (eds.), *The Boundaries of Economics*, Cambridge University Press.

—— (1981), 'Are General Equilibrium Theories Explanatory', pp. 17–32 in J. C. Pitts (ed.), *Philosophy and Economics*, Reidel, Dordrecht.

—— (1992), *The Inexact and Separate Science of Economics*, Cambridge University Press.

HAUSNER, M. (1954), 'Multi-Dimensional Utilities', in R. M. Thrall, C. H. Coombs, and R. L. Davies (eds.), *Decision Processes*, John Wiley, New York.

HEINER, R. A. (1983), 'The Origin of Predictable Behavior', *American Economic Review*, 83: 560–95.

HELM, D. (1984), 'Prediction and Causes', *Oxford Economic Papers*, 36: 118–34.

HERSHEY, J. C., and SCHOEMAKER, P. J. H. (1980), 'Risk Taking and Problem Context in the Domain of Losses', *Journal of Risk and Insurance*, 47: 111–32.

HEY, J. D. (1983), 'Whither Uncertainty?' *Economic Journal*, Conference Proceedings: 130–9.

HIRSCH, E. (1988), 'Rules for a Good Language', *Journal of Philosophy*, 85: 694–717.

JEFFREY, R. C. (1965), *The Logic of Decision*, McGraw-Hill, New York (2nd edn. 1983).

JOHNSON-LAIRD, P. N. (1983), *Mental Models*, Cambridge University Press.

—— (1970), 'Insight into a Logical Relation', *Quarterly Journal of Experimental Psychology*, 22: 49–61.

—— and WASON, P. C. (1970), 'Insight into a Logical Relation', *Quarterly Journal of Experimental Psychology*, 22: 49–61.

KAHNEMAN, D., and TVERSKY, A. (1979), 'Prospect Theory', *Econometrica*, 47: 263–91.

KANE, R. (1986), 'Principles of Reason', *Erkenntnis*, 24: 115–36.

KARMARKER, U. (1978), 'Subjectively Weighted Utility', *Organisational Behavior and Human Performance*, 21: 61–72.

—— (1974), 'The Effect of Probabilities on the Subjective Evaluation of Lotteries', Working Paper, Massachusetts Institute of Technology Sloan School of Business.

KARNI, E., and SAFRA, Z. (1987), 'Preference Reversal and the Observability of Preferences by Experimental Methods', *Econometrica*, 55: 675–85.

KATZ, D. (1960), 'The Functional Approach to the Study of Attitudes', *Public Opinion Quarterly*, 24: 163–204.

KELSEY, D. (1987), 'The Role of Information in Social Welfare Judgements', *Oxford Economic Papers*, 39: 301–17.

—— (1986), 'Choice under Partial Uncertainty', Economic Theory Discussion Paper no. 93, Cambridge University.

—— and QUIGGIN, J. (1989), 'Behind the Veil', Discussion Paper, Australian National University, Canberra.

KEYNES, J. M. (1936), *A General Theory of Employment, Interest and Money*, Macmillan, London.

—— (1921), *A Treatise on Probability*, Macmillan, London.

KIM, T., and RICHTER, M. K. (1986), 'Non-Transitive Non-Total Consumer Theory', *Journal of Economic Theory*, 38: 324–68.

KNETSCH, J. L., and SINDEN, J. A. (1984), 'Willingness to Pay and Compensation Demanded', *Quarterly Journal of Economics*, 99: 507–21.

KNIGHT, F. H. (1921), *Risk, Uncertainty and Profit*, Houghton Mifflin, Boston.

KOLMOGOROV, A. M. (1951), *Foundations of the Theory of Probability*, trans. N. Morrison, Chelsea Publishing, New York.

KOOPMAN, B. O. (1940), 'The Axioms and Algebra of Intuitive Probability', *Annals of Mathematics*, 41: 269–92.

KORHONEN, P., MOSKOWITZ, H., and WALLENIUS, J. (1990), 'Choice Behavior in Interactive Multiple-Criteria Decision Making', *Annals of Operations Research*, 23: 161–79.

KRANTZ, D. H., LUCE, R. D., SUPPER, P., and TVERSKY, A. (1971), *Foundations of Measurement*, i. Academic Press, New York.

LAPLACE, P. S. DE (1951), *A Philosophical Essay on Probability*, Dover, New York (first published 1820).

LAVALLE, I. (1990), 'New Choice Models Raise New Difficulties' in *Organisation and Decision Theory*, Kluwer Press, Dordrecht.

LEROY, S. F., and SINGELL, L. D. (1987), 'Knight on Risk and Uncertainty', *Journal of Political Economy*, 95: 909–27.

LEVI, I. (1986), 'The Paradoxes of Allais and Ellsberg', *Economics and Philosophy*, 2: 23–54.

—— (1984), *Decisions and Revisions*, Cambridge University Press.

—— (1966), 'On Potential Surprise', *Ratio*, 8: 107–29.

LEWIS, A. A. (1988), 'Balanced Games, Cores and Ultrapowers', Discussion Paper, Department of Mathematical Social Sciences, University of California at Irvine.

LEWIS, D (1986), *Counterfactuals*, Basil Blackwell, Oxford.

LINDMAN, H. R., and LYONS, J. (1978), 'Stimulus Complexity and

Choice Inconsistency among Gambles', *Organisational Behavior and Human Performance*, 21: 146–59.

LOOMES, G., and SUGDEN, R. (1982), 'Regret Theory', *Economic Journal*, 92: 805—24.

LUCE, D. (1988), 'Rank Dependent, Subjective Expected Utility Representations', *Journal of Risk and Uncertainty*, 1, 305–32.

—— (1958), 'A Probabilistic Theory of Utility', *Econometrica*, 26: 193–224.

LUCE, ROBERT, and RAIFFA, HOWARD (1957), *Games and Decisions*, Wiley, Chichester.

MACCRIMMON, K. (1968), 'Descriptive and Normative Implications of Decision Theory Postulates', in K. Borch and J. Mossin (eds.), *Risk and Uncertainty*, Macmillan, New York.

—— (1965), *An Experimental Study of the Decision Making Behavior of Business Executives*, Ph.D. thesis, University of Michigan.

—— and LARSON, S. (1979), 'Utility Theory', in M. Allais and O. Hagen (eds.), *Expected Utility Hypotheses and the Allais Paradox*, Reidel, Dordrecht.

MACHINA, M. J. (1989), 'Dynamic Consistency and Non-Expected Utility Models of Choice under Uncertainty', *Journal of Economic Literature*, 27: 1622–68.

—— (1987), 'Choice Under Uncertainty', *Journal of Economic Perspectives*, 1: 121–54.

—— (1983), 'The Economic Theory of Individual Behavior toward Risk', Center for Research on Organizational Efficiency, Stanford University.

—— (1982), 'Expected Utility without the Independence Axiom', *Econometrica*, 50: 277–323.

—— (1981), '"Rational" Decision Making versus "Rational" Decision Modelling?' Economic Theory Discussion Paper, University of Cambridge.

MAHER, P. (1986), 'The Irrelevance of Belief to Rational Action', *Erkenntnis*, 24: 363–84.

MALINVAUD, E. (1952), 'Note on von Neumann–Morgenstern Strong Independence Axiom', *Econometrica*, 20: 679.

MANKTELOW, K. I., and EVANS, J. ST B. T. (1979), 'Facilitation of Reasoning by Realism', *British Journal of Psychology*, 70, 477–88.

——and OVER, D. E. (1992), 'Mental and Deontic Reasoning', in S. Torrance and R. Spencer-Smith (eds.), *Machinations*, Ablex, Norwood, NJ.

MARSHALL, A. (1930), *Principles of Economics*, Macmillan, London.

150

MAS-COLLEL, A. (1974), 'An Equilibrium Existence Theorem without Complete or Transitive Preferences', *Journal of Mathematical Economics*, 1: 237–46.

MASKIN, E. (1979), 'Decision Making under Ignorance with Implications for Social Choice', *Theory and Decision*, 11: 319–37.

MASLOW, A. H. (1942), 'A Theory of Human Motivation', *Psychological Review*, 50: 370–96.

MAY, O. (1954), 'Intransitivity, Utility, and the Aggregation of Preference Patterns', *Econometrica*, 22: 1–13.

McCLENNEN, E. F. (1988), 'Sure-thing Doubts', in P. Gardenfors and N. E. Sahlin (eds.), *Decision Probability and Utility*, Cambridge University Press.

—— (1986), 'Dynamic Choice and Rationality', paper given to conference on 'The Foundations of Utility and Risk Theory', Aix-en-Provence, June.

—— (1982), 'Sure-Thing Doubts', paper given to conference on 'The Foundations of Utility and Risk Theory', Oslo, June (shorter version in B. P. Stigum and F. Wenstop (eds.), *Foundations of Utility and Risk Theory with Applications*, Reidel, Dordrecht).

McCORD, M., and DE NEUFVILLE, R. (1983), 'Utility Dependence on Probability', in B. Stigum and F. Wenstop (eds.), *Foundations of Utility and Risk Theory with Applications*, Reidel, Dordrecht.

McGINN, C. (1979), 'Action and its Explanation', in *Philosophical Problems in Psychology*, Methuen, London.

McGLOTHLIN, W. H. (1954), 'Stability of Choice among Uncertain Alternatives', *American Journal of Psychology*, 69: 604–15.

MEINONG, A. (1890), Review of Von Kreis's *Die Principes der Wahrscheinlichkeitsrechnung*, *Gottische Geleherte Anzeigen*, 12: 56–75.

MILL, J. S. (1863), *Utilitarianism*, pamphlet reprinted in J. S. Mill, ed. H. B. Acton, *Utilitarianism, On Liberty, and Considerations on Representative Govermnent*, J. M. Dent, London, 1972.

MINSKY, M. (1975), 'A Framework for Representing Knowledge', in P. Winston (ed.), *The Psychology of Computer Vision*, McGraw-Hill, New York.

MISES, R. VON (1957), *Mathematical Theory of Probability and Statistics*, Macmillan, London.

MONTGOMERY, H. A. (1977), 'A Study of Intransitive Preferences Using a Think-Aloud Procedure', in H. Jungermann and G. de Zeeuw (eds.), *Decision Making and Chance in Human Affairs*, Reidel, Dordrecht.

MORRISON, H. W. (1962), *Intransitivity of Paired Comparison Choices*, Ph.D. thesis, University of Michigan.

MORTON, A. (1991), *Disasters and Dilemmas*, Basil Blackwell, Oxford.

NAKAMURA, Y. (1990), 'Subjective Expected Utility with Non-Additive Probabilities on Finite State Spaces', *Journal of Economic Theory*, 51: 346–66.

NASH, J. F. (1950), 'The Bargaining Problem', *Econometrica*, 18: 155–62.

NAVARICK, D. J., and FANTINO, E. (1974), 'Stochastic Transitivity and Unidimensional Behavior Theories', *Psychological Review*, 81: 426–41.

VON NEUMANN, J., and MORGENSTERN, O. (1944), *Theory of Games and Economic Behavior*, Princeton University Press.

NG, Y. K. (1975), 'Bentham or Bergson? Finite Sensibility, Utility Functions and Social Welfare Functions', *Review of Economic Studies*, 42: 545–69.

NIETSCHE, A. (1892), 'Die Diemension der Wahrscheinlichkeit und die Evidenz der Ungewissheit', *Vierteljahrsschrift fur Wissenschaftliche Philosophie*, 16: 20–35.

NISBETT, R. E., and ROSS, L. (1980), *Human Inference*, Prentice-Hall, Englewood Cliffs, NJ.

NOZICK, R. (1969), 'Newcomb's Problem and Two Principles of Choice' in N. Rescher, Reidel, and Dordrecht, *Essays in Honor of Carl G. Hempel*, Kluwer Academic, Norwell, Mass.

PACKARD, D. J. (1982), 'Cyclical Preference Logic', *Theory and Decision*, 14: 415–26.

PETTIT, P. (1991), 'Decision Theory and Folk Psychology', in M. Bacharach and S. Hurley (eds.), *Foundations of Decision Theory*, Basil Blackwell, Oxford.

PHILLIPS, L. D., and EDWARDS, W. (1966), 'Conservatism in a Simple Probability Inference Task', *Journal of Experimental Psychology*, 72: 346–54.

POMMEREHNE, W., SCHNEIDER, F., and ZWEIFEL, P. (1982), 'Economic Theory of Choice and the Preference Reversal Phenomenon', *American Economic Review*, 72: 569–74.

POPPER, K. (1950), 'Indeterminism in Quantum Physics and in Classical Physics', *British Journal for the Philosophy of Science*, 1: 117–33.

PUTNAM, HILARY (1979), *Mathematics, Matter and Method*, Cambridge University Press.

QUIGGAN, J. (1982), 'A Theory of Anticipated Utility', *Journal of Economic Behavior and Organisation*, 3: 323–43.

QUINE, W. V. O. (1981), *Theories and Things*, Harvard University Press, Cambridge, Mass.

RAIFFA, H. (1968), *Decision Analysis*, Addison-Wesley, Reading, Mass.

RAMSEY, F. P. (1931), *The Foundations of Mathematics and Other Logical Essays*, Kegan Paul, London. (References are to the version *Philosophical Papers*, ed. David Mellor, Cambridge, University Press, 1990.)

RAPOPORT, A., and WALLSTEN, T. S. (1972), 'Individual Decision Behaviour', *Annual Review of Psychology*, 21: 131–79.

RESTLE, F., and MERRYMAN, C. (1969), 'Distance and an Illusion of Length of Line', *Journal of Experimental Psychology*, 81: 297–302.

ROBERTS, H. (1963), 'Risk, Ambiguity and the Savage Axioms', *Quarterly Journal of Economics*, 77: 327–36.

ROSE, A. M. (1957), 'A Study of Irrational Judgements', *Journal of Political Economy*, 65: 394–403.

ROSENBERG, A. (1988), *Philosophy of Social Science*, Oxford University Press.

ROTH, A. E. (1987), *Laboratory Experimentation in Economics*, Cambridge University Press.

RUBINSTEIN, A. (1989), 'The Electronic Mail Game', *American Economic Review*, 79: 385–91.

RUNDE, J. (1990), 'Keynesian Uncertainty and the Weight of Arguments', *Economics and Philosophy*, 6: 275–92.

RYLE, G. (1949), *The Concept of Mind*, Hutchinson, London.

SAMUELSON, P. A. (1952), 'Probability, Utility and the Independence Axiom', *Econometrica*, 20: 670–8.

—— (1938), 'A Note on the Pure Theory of Consumer's Behaviour', *Economica*, 5: 61–71.

SAMUELSON, W., and ZECKHAUSER, R. (1988), 'Status Quo in Decision Making', *Journal of Risk and Uncertainty*, 1: 7–59.

SARIN, R. K., and WINKLER, R. L. (1990), 'Ambiguity and Decision Modelling', Discussion Paper, Institute of Statistics and Decision Sciences, Duke University, Durham, NC.

SAVAGE, L. J. (1954), *The Foundations of Statistics*, Dover Press, New York.

SCHANK, R. C., and ABELSON, R. P. (1977), *Scripts, Plans, Goals and Understanding*, Lawrence Erlbaum, Hillsdale, NJ.

SCHEFFLER, S. (1988), *Consequentialism and its Critics*, Oxford University Press.

SCHELLING, T. (1960), *The Strategy of Conflict*, Harvard University Press, Cambridge, Mass.

SCHICK, F. (1986), 'Dutch Bookies and Money Pumps', *Journal of Philosophy*, 83: 112–19.

—— (1982), 'Under Which Descriptions?' in A. K. Sen and B. Williams (eds.), *Utilitarianism and Beyond*, Cambridge University Press.

SCHMEIDLER, D. (1989), 'Subjective Probability and Expected Utility without Additivity', *Econometrica*, 57: 571–88.

SCHOEMAKER, P. J. H. (1982), 'The Expected Utility Model', *Journal of Economic Literature*, 20: 529–63.

—— and KUNREUTHER, H. (1979), 'An Experimental Study of Insurance Decisions', *Journal of Risk and Insurance*, 46: 603–18.

SCHUMM, G. F. (1987), 'Transitivity, Preference and Indifference', *Philosophical Studies*, 52: 435–7.

SEGAL, U. (1988), 'Does the Preference Reversal Phenomenon Necessarily Contradict the Independence Axiom?', *American Economic Review*, 78: 233–6

SEN, A. K. (1987), *On Ethics and Economics*, Basil Blackwell, Oxford.

—— (1985), 'Rationality and Uncertainty', *Theory and Decision*, 18: 109–27.

—— (1979), 'Personal Utilities and Public Judgements', *Economic Journal*, 89: 537–58.

—— (1976), 'Poverty', *Econometrica*, 44: 219–31.

—— (1973), 'Behaviour and the Concept of Preference', *Economica*, 40: 241–59.

—— (1970), *Collective Choice and Social Welfare*, Holden-Day, San Francisco.

SHACKLE, G. L. S. (1961), *Decision, Order and Time*, Cambridge University Press.

—— (1952), *Expectation in Economics*, Cambridge University Press.

SHAFER, W. J. (1976), 'Equilibrium in Economics without Ordered Preferences or Free Disposal', *Journal of Mathematical Economics*, 3: 135–7.

SHEPARD, R. N. (1964), 'On Subjectively Optimum Selection among Multi-Attribute Alternatives', in M. W. Shelly and G. L. Bryan (eds.), *Human Judgements and Optimality*, John Wiley, New York.

SHIN, H. S. (1989), 'Counterfactuals and a Theory of Equilibrium in Games', mimeo, Magdalen College, Oxford.

SHORTLIFFE, E. H. (1976), *Computer-Based Medical Consultations*, Elsevier, New York.

SIMON, H. A. (1957), *Models of Man*, John Wiley, New York.

SOBEL, J. H. (1988), 'Machina and Raiffa on the Independence Axiom', *Philosophical Studies*, 54: 249–63.

SPETZLER, D. S., and STAEL VON HOLSTEIN, C.-A. (1975), 'Probability Encoding in Decision Analysis', *Management Science*, 22: 340–52.

STALNAKER, R. (1968), 'A Theory of Conditionals', in N. Rescher (ed.), *Studies in Logical Theory*, Basil Blackwell, Oxford.

STEIL, B. (1993) 'Corporate Foreign Exchange Risk Management', *Journal of Behavioral Decision Making*, 6: 1–32.

STEWART, F. (1985), *Planning to Meet Basic Needs*, Macmillan, London.

STIGLER, G. (1950), 'The Development of Utility Theory: 1 and 2', *Journal of Political Economy*, 50: 307–27 and 373–96.

STROTZ, R. H. (1956), 'Myopia and Inconsistency in Dynamic Utility Maximisation', *Review of Economic Studies*, 23: 165–80.

SUGDEN, R. (1985), 'Why be Consistent?' *Economica*, 52: 167–84.

TULLOCK, G. (1964), 'The Irrationality of Intransitivity', *Oxford Economic Papers*, 16: 401–6.

TVERSKY, A. (1975), 'A Critique of the Expected Utility Theory', *Erkenntnis*, 9: 163–93.

—— (1972), 'Elimination by Aspects', *Psychological Review*, 79: 281–91.

—— (1969), 'Intransitivity of Preferences', *Psychological Review*, 76: 31–48.

——, SATTATH, S., and SLOVIC, P. (1988), 'Contingent Weighting in Judgement and Choice', *Psychological Review*, 95: 371–84.

——, SLOVIC, P., and KAHNEMAN, D. (1990), 'The Causes of Preference Reversal', *American Economic Review*, 80: 204–17.

VILKS, A. (1992), 'A Set of Axioms for Neoclassical Economics and the Methodological Status of the Equilibrium Concept', *Economics and Philosophy*, 8: 51–82.

WASON, P. C. (1968), 'Reasoning about a Rule', *Quarterly Journal of Experimental Psychology*, 20: 273–81.

—— and SHAPIRO, D. A. (1971), 'Natural and Contrived Experience in a Reasoning Task', *Quarterly Journal of Experimental Psychology*, 23: 63–71.

WEBER, M., and CAMERER, C. (1987), 'Recent Developments in Modelling Preference under Risk', *OR Spektrum*, 9: 129–51.

VON WINTERFELT, D., and EDWARDS, W. (1986), *Decision Analysis and Behavioural Research*, Cambridge University Press.

WONG, S. (1978), *The Foundations of Paul Samuelson's Revealed Preference Theory*, Routledge & Kegan Paul, London.

VON WRIGHT, G. (1963), *The Logic of Preference*, Edinburgh University Press (2nd edn. 1991).

YAARI, M. (1965), 'Convexity in the Theory of Choice under Risk', *Quarterly Journal of Economics*, 79: 278–90.

YATES, J. F., and ZUKOWSKI, L. G. (1976), 'Characterization of Ambiguity in Decision Making', *Behavioural Science*, 21: 19–25.

ZERMELO, E. (1908), 'Untersuchungen über die Grundlagen der Mengenlehre: 1', *Math Annalen*, 65: 261–81.

—— (1912), 'Über eine Anwendung der Mengenlehre auf die Theorie des Schachspiels', *Proceedings of the Fifth International Congress of Mathematicians* (Cambridge), 2: 501–10.

Index